Molecular
CONSCIOUSNESS

"Françoise Tibika's wonderfully clear and concise book takes the reader on a step-by-step guide through the laws and processes that construct our reality. Tibika explains the complexity of our world through the living language of molecules and shows how our molecules are animated by an astonishing coherence that is the basis of our evolutionary imperative. An important read for anyone who wishes to understand how we are a part of the infinite network of information that is our universe."

KINGSLEY L. DENNIS, PH.D., AUTHOR *OF THE STRUGGLE FOR YOUR MIND* AND *NEW REVOLUTIONS FOR A SMALL PLANET* AND COAUTHOR OF *DAWN OF THE AKASHIC AGE*

"*Molecular Consciousness* begins with a very lucid and engaging account of the way that chemists currently think about molecules and the reactions they undergo. Those with a taste for speculation will be well rewarded, for the final section of the book is a much more fanciful journey into what, on one distant day, might just conceivably become science."

PETER ATKINS, AUTHOR OF *FOUR LAWS THAT DRIVE THE UNIVERSE* AND *ON BEING* AND WINNER OF THE ROYAL SOCIETY OF CHEMISTRY'S MELDOLA MEDAL

Molecular CONSCIOUSNESS

Why the Universe Is Aware of Our Presence

Françoise Tibika

Park Street Press
Rochester, Vermont • Toronto, Canada

Park Street Press
One Park Street
Rochester, Vermont 05767
www.ParkStPress.com

Text stock is SFI certified

Park Street Press is a division of Inner Traditions International

Originally published in French in 2010 by Editions le Temps Présent under the title *A quoi pensent vos molécules*

Library of Congress Cataloging-in-Publication Data
Tibika, Françoise.
[À quoi pensent vos molécules? English]
Molecular consciousness : why the universe is aware of our presence / Françoise Tibika.
pages cm
Summary: "Mind and matter are connected through information at the atomic level."—Provided by publisher.
Includes bibliographical references and index.
ISBN 978-1-59477-484-3 (pbk.) — ISBN 978-1-59477-506-2 (e-book)
1. Molecules—Philosophy. I. Title.
QD461.T5313 2012
539'.601—dc23

2012026922

Printed and bound in the United States by Lake Book Manufacturing, Inc.
The text stock is SFI certified. The Sustainable Forestry Initiative® program promotes sustainable forest management.

10 9 8 7 6 5 4 3 2 1

Text design by Jack Nichols and layout by Brian Boynton
This book was typeset in Garamond Premier Pro and Myriad Pro with AIAG used as the display typeface
Illustrations by Roy Gross

To send correspondence to the author of this book, mail a first-class letter to the author c/o Inner Traditions • Bear & Company, One Park Street, Rochester, VT 05767, and we will forward the communication, or contact the author directly at **francoise@huji.ac.il**.

To Michael

To Keren

To David

CONTENTS

PART TWO

The Secrets beyond Matter

PART THREE

When Reality Becomes a Choice

ACKNOWLEDGMENTS

My heartfelt gratitude is due to Danielle Storper-Perez—her involvement and interest gave me the impetus to write this book; to Roy Gross for his illustrations and ongoing support; to Julian E. Lipman and Charles S. Kamen for their help in proofreading the English version, and to Dr. Jerry Epstein and Rachel Epstein for their guidance.

I also thank Professors Roi Baer, Yehuda Haas, Sason S. Shaik, and Shmuel Yariv of the Hebrew University of Jerusalem for taking the time to read my drafts, as well as for their encouragement. A special thank you is due to Professor Avi Bino for his support and generosity.

And finally, I thank my friends Véronique Ayoun, Rivka Fabrikant, Gabriella Keren, Yuri Tulchinsky, and many others who invested time and energy throughout this venture, for their ready advice and encouragement.

INTRODUCTION

Science is for those who learn; poetry, for those who know.
JOSEPH ROUX (1834–1905)

Cogito ergo sum, "I think therefore I am." Can we be and not think?

The power of our mind is such that it often makes us believe we owe our existence to it. We are, however, made of molecules and atoms, and they constantly remind us that our mind doesn't have total control over our lives. Actually, we are completely subject to the demands of our molecules, which we cannot ignore without exposing ourselves to the worst dangers. Whatever the power and intricacy of our mind, we cannot exist without drinking, eating, and sleeping.

Our molecules have their own rules, and they are obviously beyond the influence of the mind. When the time is right, once the signal has been given, our molecules, still following their own rules and principles, will alter their path and turn to another "life" and, in so doing, destroy our own. They will do so whether it suits our mind or not; it will perish anyway. Our bodies will be destroyed, while our molecules, made of imperishable atoms, will join others in a new destiny.

It seems undeniable that we are made of matter and mind, and we spend much of our lives torn between the miserable needs of our flesh, made of heavy and coarse matter, and the noble aspirations of our mind, which is fine and subtle.

However, from the scientific point of view, this image is incorrect. No scientist can declare today that matter is heavy and coarse. Quantum physics teaches that matter on an atomic scale appears very different than it does on a macroscopic scale. As incredible as it might seem, in spite of all the knowledge accumulated since the beginning of civilization, from alchemy to molecular biology and the cosmology of black holes, the intrinsic nature of matter—which surrounds us and from which we are made—became, at the dawn of the twentieth century, one of the biggest enigmas of science.

Each bit of matter is an unexplainable phenomenon: a subtle and mysterious mixture, a "thing" that sometimes behaves like a particle and sometimes like a wave. Imagine, for example, something that sometimes seems like a grain of sand and sometimes looks like a ray of light. Truly wonderful! Because no better vocabulary has been found to describe this phenomenon, we refer to it as the *wave/particle duality* of matter. But this term is inaccurate. It reflects the limits of our imagination. It can also lead to confusion, because it seems to mean that matter is a wave and a particle that "cohabit," while the actual phenomenon is much more complex. Made of a countless number of these enigmatic bits of matter, what are we? A cluster of particles? A bundle of waves? Both?

On earth, we generally distinguish between living and inert matter. Living matter includes animals and plants; inert matter includes all the solids, liquids, and gases of our planet. Inert matter forms the setting, more or less immutable, in which life, unceasingly renewed, unfolds. However, chemically speaking, there are no fundamental differences between living matter and inert matter. Both are made up of the same atoms and are animated by the same energies. Indeed, the Russian biologist Alexander Oparin (1894–1980), a pioneer investigator into the origin of life, summarizing centuries of research, declared in 1924 that "the numerous attempts undertaken to discover some specific 'vital energies' resident only in organisms invariably ended in total failure."[1]

In other words, no form of energy recognized to date corresponds to that which we usually call *vital energy.* Vital energy is therefore a nonexistent concept in the scientific vocabulary. Its absence, however, did not prevent researchers from making remarkable advances in medicine. Surgical implantation of organs, artificial insemination, and the latest developments in genetics are all examples of such progress. However, in all these developments, vital energy remains outside of the equations. Vital energy is a superfluous concept in the vision modern science gives us of today's world.

The differences between a living organism and inert matter, which seem obvious to us on a macroscopic scale, quickly fade away at the atomic scale. Even the energies that animate living matter are not different from those that activate inert matter. In fact, most experts believe, as Oparin did, that living matter developed through a very long evolutionary process that began with inert matter. These first traces of life date back 500 million years to the emergence of the first bacteria.

If, for practitioners of modern science, vital energy probably belongs to the world of spirit, its source was at the core of inert matter for their predecessors, the alchemists. Back in farthest antiquity, alchemy was born from the discovery of metals and the craft of metalworking. It was a secret practice, associated with obscure experiments and transcribed in a coded language. The alchemists feared exposing themselves to Divine wrath by revealing the "science" of which they were trustees. Their goal was to discover the philosopher's stone, with which, they claimed, they could prepare the elixir of life and thereby become immortal. For alchemists, the source of wisdom and eternal life was found in the "entrails" of inert matter. Alchemists also claimed that the philosopher's stone could transform lead or mercury into gold.

Needless to say, alchemy attracted many frauds and charlatans, and was often associated with sorcery or swindling. In truth, even if the transmutation of lead or mercury into gold can be realized under very specific conditions and at very high cost, thanks to modern technologies, it is obvious that this could not have been achieved with the means available to the alchemists. This is why today alchemy does not seem very serious to us.

Yet to summarily dismiss alchemy is to forget that this science, which was born at the same time as philosophy (circa fifth century BCE), was practiced in all the great ancient civilizations, among them Egypt, Greece, and Mesopotamia. Alchemy has fascinated philosophers and scientists of all eras, such as Avicenna (980–1037), Averroes (1126–1198), Roger Bacon (1220–1292), and Paracelsus (1493–1541), as well as the great Isaac Newton (1643–1727), who devoted much of his life to it.

Employing hermetic texts composed of symbols whose meaning was hidden to laymen, true alchemists were interested in the metamorphosis of the soul. The quest of the alchemist was to study matter in order to understand his own being, thereby transcending it. The practice of alchemy constituted an inner journey, a path to supreme wisdom. Only humility and righteousness practiced over time enabled the student to walk this path, discover the philosopher's stone, and achieve metamorphosis. For that purpose, he mobilized his entire being—his fears, his guilt, his anger, his doubts, and his passions. The interpretation of an alchemical experiment had direct implications for the researcher: in purifying metals he was also purifying himself. To make gold from lead or mercury was a mystical experience and this depended more on the researcher and on his human qualities than on the extent of his knowledge. A successful experiment would result in the transformation of one's very being. From this ancient art, chemistry was born; in today's laboratories we still use utensils (test tubes, crucibles, stills) and experimental techniques (filtration, distillation, and sublimation) developed by the alchemists. However, in order to respect the limits it imposed on itself, modern chemistry had to sacrifice the mysterious aspect of its past, however rich in promise.

Did the alchemists, with their furnaces, do more than make a lot of smoke?

Can we, as the alchemists believed, metamorphose by studying matter? In the pages to come we will look deeper into this possibility. The first part of this book gives a short description of matter and of the three fundamental laws that govern it. Science calls them the three laws of thermodynamics. Our molecules are implacably subjected to them, before our birth, during our lives, and after our death. We can't escape them even for a fraction of a second.

The second part of the book briefly summarizes the discoveries that gave birth, in the twentieth century, to the quantum theory, among others. These amazing discoveries revealed our unsuspected ignorance regarding the "essence" of molecules and of all that we call *matter.*

The third part invites the reader to go beyond certain boundaries that modern science has, until now, refrained from crossing. In these zones where imagination prevails and where I intend to take you, you might perceive a closer connection between the state of your molecules and the state of your mind. In other words, you might find a bridge between your body and your mind. Will you cross it?

PART ONE

What's the Matter?

1

EVERYTHING MOVES

In 1827, the Scottish botanist Robert Brown (1773–1858) was observing particles of pollen suspended in water through his microscope. Like others before him, he noticed the constant, chaotic motion of these particles, which never stopped. This motion could not be explained by convection or by any other known physical phenomenon previously observed. Initially, Brown believed that he had finally found an expression of vital energy and saw this movement as a characteristic of the living world. However, as he continued his research, he realized that this movement occurred with all kinds of particles, whether from the living realm or not. According to his observations, the motion of any suspended particles was observable as long as the size of these particles did not exceed a few tenths (or sometimes a few hundredths) of a millimeter.

It was not until the beginning of the twentieth century that the origin of this mysterious phenomenon, now known as *Brownian motion,* was discovered. The interpretation was given by Albert Einstein (1879–1955) in 1905. Einstein is best known for his theory of relativity, but his explanation of Brownian motion was essential to the final collapse of the nonatomist theory that had reigned in science since the golden age of Greek civilization.

Einstein showed, through mathematical reasoning, that the random movement of particles suspended in water was due to collisions with even smaller particles, the water molecules themselves (which, as we will see, are made of atoms). These water molecules, which were in constant random motion, were pushing larger particles in their path like bumper cars (for instance, particles of pollen). This interpretation was revolutionary because it required the presence of atoms, the existence of which, as we will see, was still disputed at that time. For the first time, the existence of atoms was finally established. They were necessary to explain a phenomenon and their presence could no longer be denied.

Thus we conclude that matter is in constant motion whether it is alive or not. The atoms of the page that you hold in your hand are moving, as well as the atoms of the air, those of your brain, those of your heart, and those of stones. We live in a world where nothing is ever at rest. Everything is in constant agitation. It is not necessary to be equipped with a microscope to observe Brownian motion. The smoke from a cigarette, made up of tiny particles, hangs in the air and does not fall straight down like water running from a tap because the very light particles of smoke are being hit by atoms of air that are in constant motion. Their speed can be very high; at ambient temperature the speed of atoms of air can reach 600 miles per hour. However, they don't go very far. They constantly collide with each other and, on average, the distance they travel is almost zero, because their trajectory is constantly modified.

This movement of atoms is related to temperature. The higher the temperature, the faster the movement, and vice versa. When the temperature decreases, the atoms slow down and attract each other. Thus, when a gas cools, its atoms slow down and move closer to each other. Therefore, at a certain temperature, gas turns to a liquid, and at a still lower temperature it turns to a solid. In a solid atoms are very close together, but they still move. In theory, if the atoms were sufficiently cooled, they would be so tightly squeezed against each other that they would become immobilized. Calculations show that the temperature at which this would occur is −459.67°F. This temperature is called *absolute zero.* Theory tells us this temperature cannot be reached; therefore a completely motionless atom cannot exist. Above this theoretical low temperature, matter moves; the more it moves, the more its molecules are moving apart, as if the void between them was expanding like rubber, heated by the sun.

2

THE STRANGE EMPTY WORLD OF ATOMS

Chemistry teaches us that matter is composed of entities called *atoms*. The concept of the atom as the basic unit of matter (from Greek *atomos,* "that which can't be cut") goes back to the fifth century BCE. It was first suggested by Leucippus and was spread by his disciple, Democritus. Democritus proposed two principles for the formation of the universe: the "full" (atomos) and the "void" in which particles of matter move, indestructible and infinite in number, too small to be visible. This theory, which was nothing but a guess, was rejected one hundred years later, in the fourth century BCE, by the most influential of the Greek philosophers, Aristotle. For Aristotle, matter was a continuous mass; he denied the existence of a void. Because Aristotelian thought dominated natural philosophy for almost two millennia after Aristotle's death, the atomic theory, although never completely forgotten, was dismissed.

Why this great genius rejected the atomic theory, which seems quite plausible when a piece of dry bread crumbles under your fingers, will remain a mystery. Was it due to a strong intuition or a completely different reason? In fact, a modern psychologist might say that since Aristotle had been orphaned when he was twelve years old, an unfortunate event that traumatized him for life, the idea of a void terrified him. In other words, the atomic theory might have been forgotten for more than two thousand years because Aristotle feared the void. I allow myself to propose this fantasy only to point out that even behind the greatest scholars there is always a human being, with his story, his taboos, his myths, and his fears.

In any case, the existence of atoms was disputed until the beginning of the twentieth century. As mentioned earlier, it was Einstein's explanation of Brownian motion that (momentarily?) put an end to this polemic. (I write "momentarily?" because, like the atomic theory, which was never totally forgotten, the theory of the continuity of matter has also never been totally discarded).

Today we believe that matter is made up of atoms that are composed of even smaller entities: protons, with a positive charge; electrons, with a negative charge; and neutrons, which have no charge at all. We could add that electrons are fundamental particles, belonging to the lepton family, unlike protons and neutrons, which are made of quarks, but a detailed classification and characterization of such fundamental particles (the number of which increases each time a more powerful particle accelerator is built), is not necessary in the present context. We will confine ourselves to electrons, protons, and neutrons.

Atoms are so small that even the most sophisticated microscopes, which today are able to magnify a million times, do not allow us to see them. Their presence can only be detected indirectly, by observing effects they produce. We can say that the size of an atom is about a tenth of a millionth of a millimeter across; in other words, 10 million atoms lined up one next to the other would fit in the length of one millimeter.

In the center of the atom, the protons and the neutrons, bound together, form the atomic nucleus. Given the laws of electromagnetism, which stipulate that identical charges repulse each other, this raises the question of how protons stay together in the center of the atom despite their identical charges. In fact, there is a force much more powerful than the electromagnetic force. It is called the *strong force.* It is not found anywhere other than in the atomic nucleus. The strength of this force is approximately a hundred times greater than the force of repulsion between the protons, so it is able to hold them together.

The strong force and the electromagnetic force are two of four forces acting on matter that are recognized today. The other two are gravity and what is called the *weak force.* These four forces represent the four different kinds of "glue" that nature uses to construct this enormous edifice we call the universe. They "weld" matter together and prevent our universe from falling apart. Without them, nothing would cohere, and our whole world would just be fine dust, lost in the cosmic void. Gravity sticks us to the earth and prevents us from leaving the ground, being swept up by the winds. It makes the earth turn on its axis and maintains it in its orbit; it also makes all the planets of all the galaxies turn and makes our cosmos look like a well-lubricated system of wheels. The weak force is much less familiar to us than the other three because we don't experience it directly in our everyday life. The weak force holds together particles that separate only during certain nuclear reactions. It is called the weak force because, although of nuclear origin, it is much weaker than the strong force.

These four forces are responsible for the integrity of our universe. Of them, the strong force is by far the most powerful. Its intensity is evoked by the extraordinary density of the atomic nucleus, nearly 200 million tons per cubic centimeter. The heart of the atom thus contains energies that make us tremble today: their manipulation enabled the construction of the atomic bomb.

Around these fantastic and puzzling nuclei, extremely light electrons are whirling, nearly as fast as the speed of light, 186,000 miles per second. Each electron is approximately two thousand times lighter than a proton or a neutron, and much smaller. If protons and neutrons were small marbles one centimeter in diameter, the diameter of an electron would be smaller than a dust mote. An atom would be a few kilometers in diameter, that is, about one thousand million times bigger than an electron. We can picture it this way: the distance between the nucleus and the edge of an atom, from the electron's perspective, is greater than the distance between the earth and the moon from our perspective. In fact, the space inside the atom in which the electrons are whirling is so large, compared to their size and their number, that 99.999999999999% of the atom is empty.

In other words, if we keep in mind that we are made of atoms, we reach the astonishing conclusion that we are practically empty. Indeed, if our atoms could be compressed and the void between the electrons and their nuclei removed, we would be so small that an electronic microscope would be needed in order to see us. We could stack all the population of the globe in one cubic centimeter and there would still be room. In fact, if all of the void of the universe could be removed, we might be in the same situation as before the Big Bang; that is, the whole universe would contract to a point.

In summary, questioned by modern science, matter reveals to us that we all are victims of a stunning optical illusion, that almost all the mass of the universe is contained in very dense nuclei and that the rest, that is, 99.999999999999% of the volume of our universe, including the volume of our own body, is empty. Therefore, if someone tells you one day that you are insignificant, don't get annoyed: it is factually true, but you are not the only one! Nevertheless, within the void that is you, there are billions of billions of billions of billions of billions of atoms, like tiny galaxies.

For most of us, most of the time, this strange world doesn't enter our daily thoughts, and yet this world is ours. It is that of our surroundings and our own flesh, the only one within our reach—the one that controls the simplest of our gestures, the slightest of our shivers.

3

THE ELEMENTS, THE ONLY INHABITANTS ON EARTH

In the fourth century BCE, the Greek philosopher Empedocles proposed that all matter was a combination of four elements: fire, water, air, and earth. Aristotle liked this theory and adopted it. Thus, it dominated Western thought until the dawn of modern science, when it was abandoned completely.

Modern science today sees matter not as a combination of fire, water, air, and earth, but, rather, as Democritus had imagined, as composed of atoms. All atoms are not the same. In fact, there are more than one hundred types of atoms, of different composition.

These hundred types of atoms have become the "new" elements of our universe, those of which the entire world is built. An element is a piece of matter containing one kind of atom. Each time an element was discovered, it got a name and a symbol. These elements have been carefully ordered in a table, the periodic table (see figure 3.1).

All matter is a combination, more or less complex, but always different, of these elements—sometimes of a few elements, sometimes of many. The number of possible combinations of these elements seems unlimited.

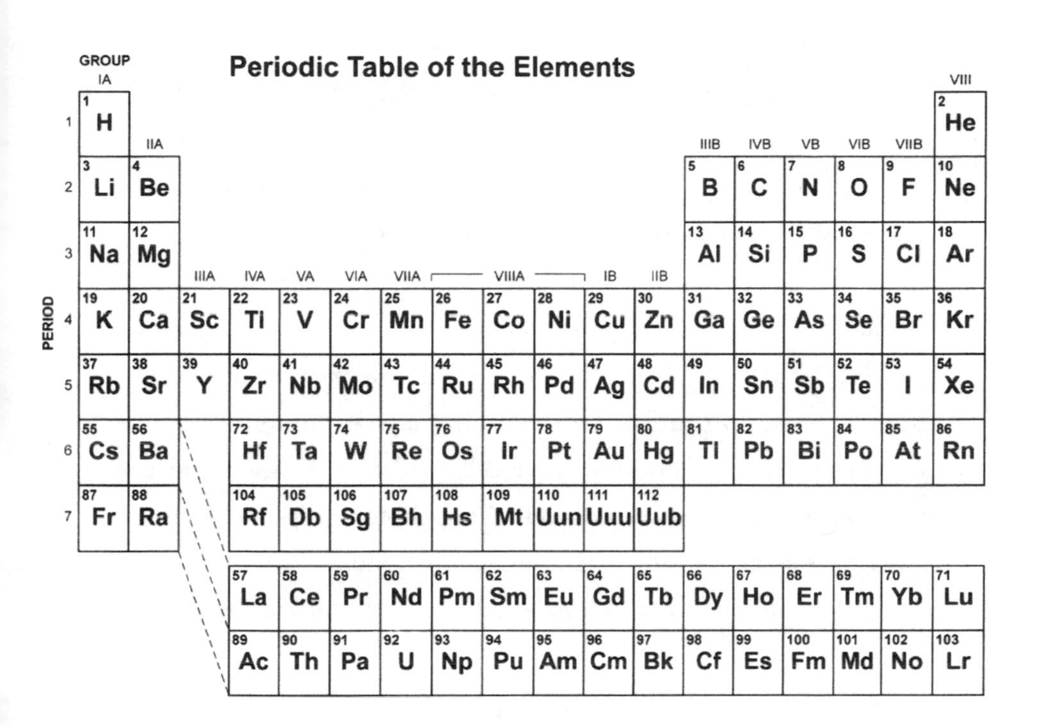

Figure 3.1. The periodic table, also called Mendeleev's Table, after the nineteenth-century Russian chemist Dimitri Mendeleev (1834–1907), who was largely responsible for its conception.

The periodic table can be viewed as a summary of the census of all the "inhabitants" of our planet and even, until proven otherwise, of all the "inhabitants" of our universe. So far, it seems there is no one else on our planet other than these elements. Everything we see is made of them. A handful of them are artificial, like berkelium (Bk) or einsteinium (Es); in general, these can be found only in the laboratories where they were created. Most of them are quite unstable and their existence is sometimes limited to a fraction of a second, after which they decompose. All the other elements of the periodic table are found in nature. Some are very familiar to us, like calcium (Ca), sodium (Na), copper (Cu), and phosphorus (P), and some are very rare, like lutetium (Lu) or samarium (Sm).

Earth's crust consists mostly of oxygen (O) and silicon (Si). Living matter is made up mostly of oxygen (O), carbon (C), hydrogen (H), and nitrogen (N). These four elements represent more than ninety-five percent of our weight. Many other elements are as essential to life, but generally they are found only in very small quantities in our organism. However, if they exceed a certain concentration, they can be fatal. Cobalt (Co), for example, is part of vitamin B_{12}, essential to our organism. This vitamin plays an important part in the normal operation of our nerve cells and also in the production of DNA. The body of an adult has about three milligrams of cobalt, and a person cannot safely accumulate much more. Exposure to high levels of cobalt can result in lung and heart deficiencies, among other conditions.

The elements differ from each other in the number of their protons. As this number is specific to each element, it is like an identity number. In the periodic table of fig. 3.1, this number is written at the top left corner of each symbol; elements are sorted by ascending order of the number of their protons. Curiously, from one element to the next, this number increases regularly by one unit. It is the only parameter that can identify an element beyond any doubt. The atom of an element may lose or gain electrons, or it may lose or gain neutrons, and remain the same element, but if it loses or gains a proton, it will be transformed into another element. For instance, gold (Au) is an element that contains seventy-nine protons; lead (Pb) has eighty-two. In theory, the transmutation of one element to another is quite easy. By removing three protons from an atom of lead, one could obtain an atom of gold. But any manipulation that affects the atomic nucleus requires such a large investment of energy that, even supposing it were technically possible, which is rarely the case, the cost would far exceed the anticipated return. Still, we cannot blame alchemists for following a totally false intuition, thinking it was possible to turn lead into gold.

4

THE MAGICAL POWER OF CHEMICAL BONDS

As we stated previously, there is no one else on our planet other than these elements. All matter is made up of these elements and the great majority of them are "social beings," that is, one finds them bound together. The links between atoms are known as *chemical bonds.* In general, atoms are bound together in groups of two or more. This is true for all but six elements of the periodic table. The atoms of these six elements don't like company; they bind with great difficulty and are almost always found alone. These elements are gases at room temperature: helium (He), neon (Ne), argon (Ar), krypton (Kr), xenon (Xe), and radon (Rn). They could have been called wild, or snobs, from Latin *sine nobilitas,* meaning "without nobility," but on the contrary the chemists called them the noble gases. They appear in the extreme right-hand column of the periodic table.

The atoms of all the other elements are found in nature, bound together. Atoms bond either to atoms of their kind—as with many gases we know, such as oxygen, nitrogen, and hydrogen—or to atoms of different kinds, such as in water. In these bonds, electrons seem to have the more prominent role. They "glue" atoms together. There are several modes of binding, but all are the result of the attraction between the electrons of one atom and the nucleus of another. This attraction of an electron to a nucleus other than that of "its actual atom" forces both atoms to come closer, and this special proximity creates a bond, a chemical bond. To get an idea of this proximity, consider a noble gas like helium (He): the distance between two helium atoms at room temperature and pressure is 100,000 times the distance between two bound oxygen atoms forming oxygen gas (O_2). In fact, chemical bonds are the fruit of the electromagnetic force. This force pushes negatively charged electrons toward positively charged protons.

Atoms held together by chemical bonds form combinations called *molecules*. Matter is made of molecules. If all the molecules of a piece of matter are identical, that matter is called a *pure compound*. Water is a pure compound; it is made only of water molecules. In a water molecule one oxygen atom, O, is bound to two hydrogen atoms, H (see figure 4.1). Its chemical formula is H_2O. The chemical formula

Figure 4.1. The structure of a water molecule, H_2O

for a molecule corresponds to its atomic composition. When there are two or more atoms of the same element in a molecule, this number is written on the right side of the element's symbol.

In contrast there are also compounds that are not pure. Wine, for example, is not a pure compound, because it contains molecules of different kinds: alcohol, sugar, water, and so on.

Some molecules contain very few atoms, and sometimes only two, such as oxygen (O_2), while other molecules, such as proteins, contain hundreds of thousands of atoms (see figure 4.2).

We cannot see atoms, even under the microscope, and our drawings of them are partly the fruit of our imagination. However,

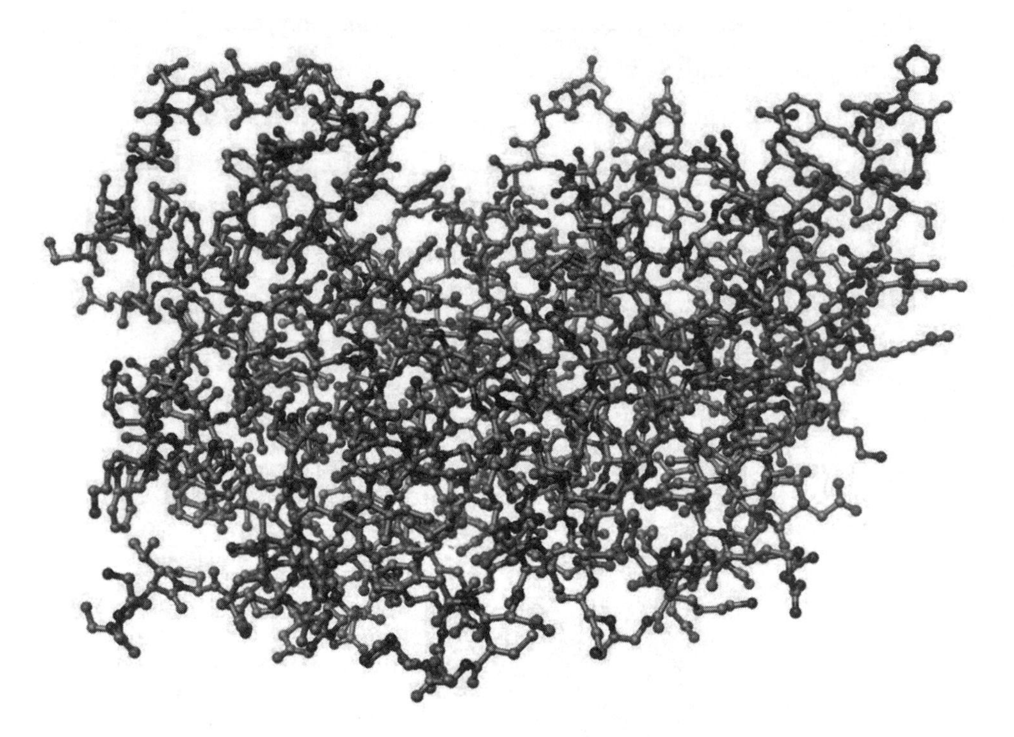

Figure 4.2. A computerized image of a protein, xylanase, extracted from the bacteria Geobacillus stearothermophilus *(courtesy of Gil Shoham and Vered Solomon from the Hebrew University)*

using a very complex analytical procedure, we can deduce the internal geometry of the molecules of a compound with a high degree of accuracy. For example, we can calculate the average distance and the average angle between atoms. This information is very important to chemists, because the properties of a compound depend greatly on the arrangement of its molecules. For example, the atoms in a water molecule are not arranged in a straight line, but form an angle of 104 degrees with each other. Were it not for this angle, that is, if water molecules were linear, the boiling point of water might be much lower than 212°F, meaning that water could evaporate at ambient temperatures. In other words, without this angle, we would not be here today.

A fuller discussion of these chemical bonds would exceed the objective of this book. The important point regarding our subject is the almost magical power of these chemical bonds in holding atoms together. They have a dramatic effect on the elements, transforming them in ways that can make them unrecognizable. The elements can easily change texture, color, odor, and so on, and acquire new properties. The various facets that each element of the periodic table exhibits by virtue of these bonds are astonishing, as much by their number as by their diversity.

For example, nitrogen, N. Nitrogen is the name of an element, but also the name of the gas composed of molecules containing two atoms of nitrogen bound together, N_2. The air we breathe contains 21 percent oxygen, O_2; almost all the rest, that is, close to 80 percent, is nitrogen, N_2. In other words, we breathe oxygen diluted in nitrogen. Nitrogen gas is completely colorless, odorless, and inoffensive, and very stable. Its stability is the reason why it is inoffensive to us. Nevertheless, if its concentration were to increase, we would suffocate.

Nitrogen gas, N_2, decomposes with great difficulty because the chemical bond that exists between two nitrogen atoms is very strong. However, isolated nitrogen atoms are found in a countless number of combinations, each very different from the other. For example, with hydrogen, nitrogen can produce ammonia, a very dangerous and noxious gas, or hydrazine, an explosive liquid that is used as fuel in the space shuttle. Combined with oxygen, nitrogen can produce nitrous oxide, a colorless, euphoric gas also called laughing gas, that is used as an anesthetic, or nitrogen dioxide, a brownish-red, corrosive, and toxic gas. Nitrogen is also found in the white and odorless nitrate salts, the main components of chemical fertilizers. Nitrogen is also a major component of our own body; it is found in all the proteins that compose our tissues, from that of our aortic valve to the nail of our little toe.

A human body contains 2.4 percent nitrogen, which is not a small amount if one considers that 60 to 80 percent of our weight is water. In fact, in an adult weighing 150 pounds there are about three and a half pounds of nitrogen. The nitrogen in our tissues is renewed thanks to the proteins we absorb through the foods we eat. Its source is the fertilizers used to grow crops, which in turn are eaten by farm animals. In one way or another, the nitrogen in these fertilizers finally arrives, directly or not, on our plate. It is striking to think that without these nitrate salts there might not have been enough nitrogen for us to manufacture our proteins. One thing we know is that there is not enough for everybody, or at least it is not equally shared.

Now for carbon, C. Bound to other atoms, it is found in an enormous number of chemical compounds: it is in carbon dioxide (CO_2), a gas essential to the balance of our atmosphere, which we exhale and which plants absorb during photosynthesis; it is also in a very dangerous gas, carbon monoxide (CO). Not only is carbon monoxide lethal, but it is very easily generated—in a stove, for example, or a badly ventilated chimney. Emile Zola (1840–1902), the French novelist, died of carbon monoxide poisoning because of a blocked chimney. Another small oxygen atom in the gas molecules and Zola might have been saved! Carbon also appears in an impressive number of other gases, oils, and waxes; in acetone, sugar, cotton, alcohol, wood, and gasoline. In humans, carbon constitutes about 10 percent of our weight. Carbon atoms belong to the very structure of our organism, more than any other element. They appear in our body as very long chains to which are attached, in one way or another, most of the other atoms crucial for our survival.

Any chemical compound can metamorphose as soon as the kind, the number, or even the position of its atoms is modified. Its shape, its color, its smell, and all its other properties might be strongly affected. The example of graphite and diamond is particularly spectacular. They are the only natural compounds composed of pure carbon. Both contain only carbon atoms, but the atoms in graphite and in diamond are arranged differently. In graphite, each carbon atom is bound to three other carbon atoms, and in diamond to four others. At the molecular level, the result is that graphite is made of separate layers of carbon atoms, one on top of the other like phyllo dough, with no true chemical bonds between these layers, as shown in figure 4.3, while diamond is a three-dimensional network of carbon atoms all bound to each other, as shown in figure 4.4 (see page 26).

At the macroscopic level, this structural difference has dramatic repercussions on the appearance and the properties of these two materials. Graphite is dark, opaque, and brittle, whereas diamond is transparent, bright, and very hard. However, if diamond is heated to nearly 1000°F for few minutes its atoms become mobile enough to shift: the diamond turns to graphite, that is, into ash. Moreover, this transformation is reversible: under very high pressure, these carbon ashes may be transformed back into diamond.

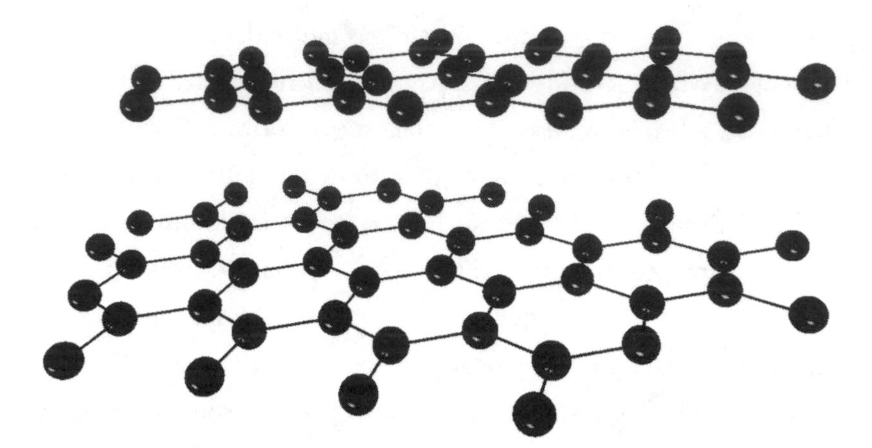

Figure 4.3. Computerized image of graphite layers

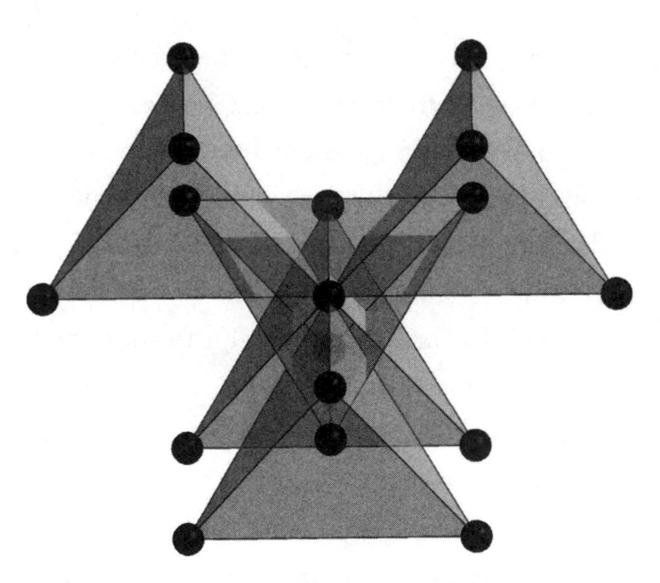

Figure 4.4. Computerized image of the structure of diamond (drawn by Shmuel Cohen from the Hebrew University)

Chlorine (Cl) can be found in bleach but also in cooking salt, plastic can be manufactured from petroleum, a tiny seed can be transformed into a baobab tree. Likewise, an egg can turn into an omelet or a chicken, depending on its circumstances.

In short, thanks to the astonishing property of elements to metamorphose through their chemical bonds, they reveal their many facets, which depend on their bonds. Therefore only a hundred elements are enough to generate all the solids, liquids, and gases of our planet; all the variety of its scenery; the earth, sky, plants, and animals; and all the synthetic products that chemists have already succeeded in manufacturing, the list of which lengthens every day. In short, with only the elements of the periodic table, it is possible to manufacture all that is around us and also all that is in us.

Indeed, what are we, other than a pile of chemical compounds in constant motion? A walking chemical factory that, while consuming energy and expelling waste, is busy day and night creating and breaking down an amazing number of chemicals. In fact, our organism is occupied with only one activity: the creation and breaking of chemical bonds. This is all it can do. Undeniably, this activity also transforms us: we grow taller, older, fatter; the color of our skin and of our hair changes; and so on.

The creation, break, or any modification of a chemical bond is called a *chemical reaction*. This is the subject of the next chapter.

5

CHEMICAL REACTIONS

The Psychology of Elements

In the previous chapter we discussed how the electrons in a chemical bond serve as glue between the atoms. However, as atoms and electrons move unceasingly, this glue doesn't look like dry glue, but like fresh glue that is moving. This movement is quite regular, even rhythmic, which is why a chemical bond is compared to a spring connecting two atoms, a spring that can be stretched, compressed, and even twisted.

In a very simple way, one may say that, depending on the atoms they bind and on the surrounding conditions, these springs can break. The breaking of a spring or, in other words, the breaking of a chemical bond between two atoms, is triggered by their neighbors, together with the temperature, the pressure, the light, and so on. When a bond breaks, it is because electrons feel a stronger attraction toward their neighbors than toward the nuclei around which, so far, they had been orbiting. Usually bonds between atoms break to form new ones. At this point, there is an exchange of partners and a reorganization of the chemical bonds. The rupture of a chemical bond or the formation of a new one is a *chemical reaction.*

This exchange of partners is not systematic; mixing different molecules does not guarantee that the electron of one molecule will be attracted to the nucleus of the other molecule and that a reaction will occur. In fact, the elements have particular affinities toward one another; they form bonds with certain atoms and not with others. Some of these bonds are very strong and others are quite weak. Each element has a particular way of reacting with any other element. The elements have a "character" that, like our own, is reflected in the nature and stability of their bonds. Let's look at two examples of the manifestation of their character.

Like vitamins, some minerals are essential to our organism. Iron (Fe) is surely one of the best known. Its paramount role is to serve as a "taxi" for oxygen (O_2). Once inhaled, oxygen is absorbed into our blood by simple diffusion. There it binds to the iron contained in our blood, and the iron transports this oxygen to all our cells as a taxi. At the right place and time, the bond between iron and oxygen breaks, the oxygen leaves the taxi and is discharged into our cells. There it triggers a complex system of intricate reactions, at the end of which the energy necessary to the good functioning of our cells is released. Thanks to the oxygen we constantly breathe, chemical energy is constantly being produced and used by our body.

A manganese atom (Mn) has only one proton more than an iron atom, but if iron were replaced by manganese, we would die very quickly. Oxygen binds to manganese, but oxygen's bond to manganese is different from the one it forms with iron; manganese would not release the oxygen on time. Oxygen would not reach its destination, and you can imagine the rest. No other element can replace iron in this task.

Before moving to the second example, we should note that the production of energy by our cells, which oxygen makes possible, is accompanied by the production of a gas, carbon dioxide (CO_2), which we get rid of each time we exhale. We constantly inhale oxygen and exhale carbon dioxide. Depriving an individual of this exchange with the outside for just a few minutes will disturb the whole organism and puts an end to its life. It is curious to note that a carbon dioxide molecule looks like an oxygen (O_2) molecule with a carbon atom (C) grafted onto it (see figure 5.1)—as if, with each oxygen molecule that we breathe, we are "taxed" with a carbon atom.

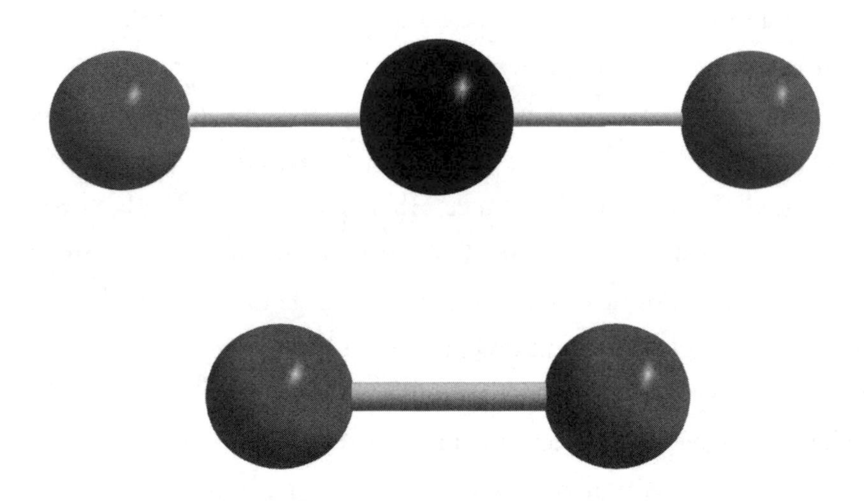

Figure 5.1. In this computerized image of a molecule of O_2 (below) and a molecule of CO_2 (above), oxygen atoms are in gray and the carbon atom is in black.

In quite another context, the production of energy by taking in oxygen and releasing carbon dioxide is known as *combustion*. For example, wood is a mixture of chemicals, primarily composed of carbon. Like all fuel, it burns only in the presence of oxygen and, during its combustion, heat and carbon dioxide are released. Therefore breathing—where O_2 is taken in, CO_2 released, and energy produced—can be regarded as combustion of our carbon reserves without fire or flames. Thus there is nothing unique about the energy that our body produces, which we call our vital energy; its production can be compared to the release of energy by a piece of burning wood.

Now to our second example of how elements manifest their character. As already mentioned, nitrogen is an element, but also a gas, N_2, composed of molecules containing two atoms of nitrogen, one of the most stable existing compounds. The bond between these two nitrogen atoms is very strong; it cannot be broken unless the atoms are heated to 900°F and subjected to a pressure of at least 300 atmospheres. In nature, the energy of a thunderstorm or the ingenuity of quite specific bacteria is needed to break this bond between two nitrogen atoms.

Figure 5.2. The atoms of nitrogen in nitrogen gas are well bonded.

If the bond between two nitrogen atoms is difficult to break, the reaction involved in forming this bond is sometimes very loud and spectacular. The attraction between two nitrogen atoms, each of them bound to another atom, can sometimes cause a violent reaction. The nitrogen atoms separate violently from their former partners and bind together. The energy released in this process is so great that the motion of the surrounding molecules is accelerated until they exceed the speed of sound (768 miles per hour), causing a thunderous noise and much damage along the way. As you may have guessed, this reaction generates an explosion. In fact, very often, an explosive is only a chemical system that, under certain conditions, triggers the meeting of two formerly separated nitrogen atoms. The meeting of two oxygen atoms, to generate O_2, for example, is much less spectacular.

Undoubtedly the strength of the bond between two nitrogen atoms is remarkable. However, metallic lithium (Li) can break it at room temperature. Indeed after exposure to air, lithium is covered with a thin layer of a mixture of Li and N. Lithium is the only element of the whole periodic table whose presence can cause the rupture of the bond between two nitrogen atoms at ambient temperature. It should be noted, however, that the formation of the bond between Li and N is not spectacular at all.

Thus elements are unique in the way they react. They have their "preferences" and their "antipathies." In other words, they each have their own character. There are, however, similarities among the characteristics of various elements, and these similarities allow them to be grouped into "families."

In 1869, Russian chemist Dimitri Mendeleev published the first classification of the elements, which he revised in 1871. This classification was the initial version of what became the famous periodic table, also called the Mendeleev table in his honor. In this table the elements in a column are members of the same family and often behave in a similar manner. In Mendeleev's time, not all the elements of the periodic table had been isolated and labeled. Mendeleev inventoried and classified the known elements according to their weight and other physical and chemical properties.

His genius in constructing the table was the inspiration to leave empty slots. He realized that, according to the criteria of classification that he had established, no elements then known occupied those slots. They were reserved for still-unknown elements. He predicted their existence and even some of their properties. He was fortunate that some of them were discovered while he was still alive: scandium (Sc) in 1875, gallium (Ga) in 1878, and germanium (Ge) in 1886. These elements had the characteristics that Mendeleev had predicted according to their positions in the table he had designed. He died covered with honors, and element 101 of the periodic table, mendelevium, bears his name.

The periodic table shows that oxygen (O) and sulfur (S) belong to the same family. They are in the same column, one above the other. They often form bonds of the same kind and with the same elements. Their compounds are made of molecules that look quite similar—for example, water, H_2O, and hydrogen sulfide, H_2S. However, we have to be careful in drawing these comparisons. Water is an odorless liquid vital to this planet, whereas hydrogen sulfide is not only a foul-smelling gas but also a deadly poison!

Books have been written about each of the elements, their bonds, their reactions, and the properties of their compounds. These books are constantly being updated, because new discoveries are continually being made about them. Each element seems to have an individual character that is so rich it appears inexhaustible and, at times, as unexpected as our own characters; every day, throughout the world, researchers discover new behaviors and new kinds of bonds hitherto unsuspected.

For this reason, chemistry is to matter what psychology is to humanity. It studies the character of these surprising elements and their reactions. It studies how and under what conditions an element reacts as it does, why under certain conditions its electrons are attracted to some nuclei more than to others, and why some bonds are stronger than others.

Thanks to the knowledge already accumulated, chemists have become specialists in discovering the elements' mechanisms and in their manipulation. Their goal: to create new combinations and manufacture new compounds having new properties. These are usually called *chemical* products, but since every product is chemical, they should be called *synthetic* products. The chemists' success is impressive: 90 million compounds have already been invented and thousands of others are added each week! Some play an integral part in our daily life: drugs, detergents, dyes, foodstuffs, plastics, and many other new materials that modify our life almost daily.

In summary: elements bind selectively; certain bonds can be created, whereas others cannot; certain bonds are quite strong, and others much weaker. However, their stability is relative; it always depends on the other alternatives. In other words, environment has a major influence on chemical bonds.

At the end of the nineteenth century it was discovered that not only did the elements each have their individual character, but that the reactivity of their molecules was governed by very rigorous principles, the *three laws of thermodynamics*. Like a social code, these laws prescribe the behavior of atoms and molecules, whose "social lives" are quite enviable in their simplicity: in the world of atoms, exchange of partners is permitted only within the strict framework of these three laws. Every atom obeys—no transgressors!

To simplify, let's call them the *three laws of matter.* They apply to inert matter as well as to living matter. Even though they are unknown to most of us, these laws govern all that is around us. They control the behavior of our molecules and dictate the course of the simplest of our reactions. Although we are able to build sophisticated machines with which we can temporarily overcome the laws of gravity, we can never escape these molecular laws. Indeed, how could we escape ourselves?

As all of our molecules are ruled by these laws, we might wonder what purpose our mind serves.

6

THE FIRST LAW

We Cannot Make or Destroy Energy

The first law of thermodynamics is undoubtedly the best-known of the three laws, and it relates to energy. Energy corresponds to the capacity of a system to modify another system. Having energy means having the ability to do lots of things. To do things means to modify a portion of the molecular soup in which we dwell. When we have energy, we often spend it without much thought because we know we will probably replenish it, primarily by eating, drinking, and sleeping. We draw our energy from outside. But are we able to manufacture energy?

With all our intelligence and power, we do not have the ability to create energy for ourselves or for anybody else. Though we often speak about the production or loss of energy, this is an abuse of language. We can use energy, transform it, or even store it, but we cannot create it. We can manipulate energy, if only up to a certain extent, but we cannot make energy. This is what the first law of thermodynamics stipulates: energy can never be created; it also says that energy can never be destroyed.

Nature is our only source of energy. This gift of nature is our inheritance and our true wealth. Without energy a system has no impact on the world; its presence, if this word has meaning in such a circumstance, goes unnoticed. Energy is provided by wind, rain, oceans, rivers, plants, and animals, but mostly by the sun, its heat and its light. Without the heat of the sun, the temperature of Earth would be hundreds of degrees lower, and there would be no life on this planet. We are unable to produce energy, just as we are unable to produce a baby in a laboratory. There is no synthetic or artificial energy. On the other hand, energy is imperishable and indestructible.

It seems that during the Big Bang, all the energy of the world was allocated once and for all. Since then, its amount has neither increased nor diminished, nor can it disappear. However, even though energy is indestructible, it can move from one system to another. In other words, it circulates, rather like a fluid. While a fluid changes its form according to its container, energy changes its form according to the system that contains it. For example, the energy contained in food is chemical energy; in a battery, electric energy; in the wind, mechanical energy; and so on. When you eat, you convert the chemical energy that you draw from food into thermal energ, that is, heat, to maintain the temperature of your body at 98.6°F, or into mechanical energy to move your muscles, and so on.

In a similar way, the hydraulic energy of a waterfall can be transformed into electric power, the chemical energy of oil into thermal energy, the electric energy of a battery into mechanical energy, and so on. It is worth noting that these transformations from one form of energy to another may be carried out directly or by stages. For example, for hydraulic power to be transformed into electric power, it must first be transformed into mechanical energy to make turbines turn, and only then can this mechanical energy be transformed into electric power. But, in theory at least, energy can take nearly any form.

The first statement of this law—"Energy can be neither created nor destroyed; it can only transform"—was attributed to a German doctor, Julius Robert von Mayer (1814–1878) in 1841. One year earlier, while he had been in Indonesia, near the equator, he noticed that blood from the veins of sailors was redder than that of his patients in Germany. Mayer established a link between the color of the blood and its oxygen (O_2) content. The redder the blood, the more oxygen it contained. At that time it was already widely accepted that oxygen was used in the production of the energy of living organisms. Mayer deduced that there was an excess of oxygen in the blood of the sailors because they produced less energy. They produced less energy because they needed less energy. And they needed less energy simply because the weather in Djakarta is hotter than in Tübingen. When it is hot outside, obviously less energy is needed to maintain the body at 98.6°F. Therefore, in the blood of the people from Indonesia, there is more oxygen than in the blood of Germans. Mayer discovered the connection between thermal and chemical energy: these energies were mutually convertible.

If we imagine our body as a fantastic chemical factory, and then consider the electricity bill for the energy consumed to keep that factory working, we would have to pay more or less the same amount each month. In other words, our personal need for energy is more or less constant. When our consumption of energy fluctuates too greatly, our organism is disturbed. If we consume more energy than usual, we grow fatter, and when we do not consume sufficient energy, our organism weakens. It weakens because it must draw its energy from its own reserves: first, the (chemical) energy contained in fat, then in the muscles, and so on.

Although our energy needs are constant from adulthood to old age, the energy takes various forms and is used for various tasks. We use energy to heat our body and maintain it at a temperature of 98.6°F, like a stew that simmers. (Actually, if it varies more than three or four degrees, this stew decomposes and starts to rot and stink.) We also use energy to move our body: our legs, our arms, our hands, our fingers, to open our mouth and our eyes. Undeniably, part of this energy is also invested in our mental activity. It is tiring to concentrate for a long time at the office or at school, as is playing on a computer or playing chess; even lying in bed ruminating about our sorrows and our torments is exhausting.

How is your energy distributed? How much is used to maintain your body temperature? What percentage is invested in motion? In sport? In studies? In anger? In hate? In joy? Who controls it? If we could not control the form our energy takes, we would not be very different from a vegetable. We have a mind of our own and it seems that at this stage it has a statement to make: "Take advantage of this opportunity, because it might not be available elsewhere." In this world dominated by the laws of matter our mind has only one lifetime and limited power. This power could just be to have a certain control of our energy.

Thus, energy is convertible. It takes various forms but remains indestructible. How, then, can we speak about a world energy crisis, if energy is indestructible? The answer is in the next chapter, on the second law of matter.

But before closing this chapter, let's consider one other facet of the law of the conservation of energy: neither energy nor matter can disappear. Indeed, can you make an atom disappear? They are so small and so numerous that you might be tempted to say yes. But how would you proceed? Would you break it? Would you hide it under the pillow? Would you launch it into the sky? Would you bury it in the ground?

Once again, Einstein gives us the answer. His famous formula,

$E = mc^2$ (E = energy, m = mass, c = speed of light) stipulates that if a mass, m, disappears from the universe, its departure would not go unnoticed; in fact, this would create an enormous explosion and a tremendous amount of energy—equal to mc^2—would be released. In other words, an energy equivalent to the mass, m, multiplied by the square of the speed of light, c^2. In fact, any mass is a reservoir of energy, a special reservoir because it disappears while being emptied of its contents. It is as if matter was condensed energy, leaving nothing else after it explodes than energy.

Einstein's formula stipulates, for example, that if you were able to make a grain of salt disappear, the energy released would be enough to destroy an entire building. The energy contained in an adult weighing 150 pounds would be the equivalent of 1 billion tons of TNT. (By comparison, the first atomic bomb dropped on Hiroshima on August 6, 1945, exploded with an energy equivalent to 20,000 tons of TNT.) This energy is contained in the nuclei of our atoms. It can be released only if all our atoms explode and disintegrate; as you might imagine, though, this won't happen tomorrow. The disintegration of atoms is observed only in the cosmos or in specialized laboratories, during the manipulation of certain radioactive materials. The quantity of energy released during these reactions is what Einstein's formula predicts.

The law of conservation of matter was already known at the time of Von Mayer. It was expressed by this famous sentence, often attributed to the French chemist Antoine Laurent Lavoisier (1743–1794): "Nothing is lost, nothing is created, all is transformed." Lavoisier is often regarded as the father of modern chemistry, as he was the first to introduce the systematic use of scales into all his experiments. This is how, by the end of the eighteenth century, the spirits and demons that haunted alchemy were gradually transformed into grams and milligrams.

Lavoisier kicked out one particular monster that had haunted laboratories for nearly a century. At the end of the seventeenth century, it was thought that all inflammable materials contained an odorless and colorless substance, phlogiston. It was thought that when heat was applied to a material, its phlogiston was released and the material was then left in its elementary form, without phlogiston, resulting in a loss of weight—as, for instance, when wood burns and is transformed into ashes, that is, wood without phlogiston. Belief in the phlogiston theory began to weaken after Lavoisier introduced the scale as an essential experimental tool. That enabled him to show that sometimes, against all expectations, when phlogiston left a compound, its weight *increased*.

The most enthusiastic adherents of the phlogiston theory declared that, in certain cases, the phlogiston had a negative weight, which explained the increase in the compound's weight after the phlogiston had left it—for example, when magnesium (Mg) burns. Lavoisier showed that combustion was due to a reaction with oxygen in the air, and that certain metals, such as magnesium, became coated with an oxide layer during their combustion, which increased their weight. The phlogiston theory paled in comparison to Lavoisier's brilliant observations and was finally abandoned.

Unfortunately, despite his genius, Lavoisier was condemned to death by a French revolutionary tribunal and guillotined in 1794 at the age of 51, after a trial that lasted less than one day. The great French mathematician Joseph Louis Lagrange (1736–1813) is said to have remarked shortly after the execution, *"Il ne leur a fallu qu'un moment pour faire tomber cette tête, et cent années ne suffiront pas pour en reproduire une semblable."* ("It took them only an instant to cut off his head, but a hundred years will not be enough to produce another like it.")

If the first law of matter—"Nothing is lost, nothing is created, all is transformed"—did not hold, we would never be sure that we would be able to find lost keys or even that the bed in which we went to sleep would still be there when we woke up in the morning! I am telling you these fantasies to introduce the second law of matter and to make you realize that even if this second law is lesser known than the first, it is no less fundamental.

7

THE SECOND LAW

Chaos Rules

At the end of the eighteenth century, energy was recognized as being indestructible and convertible. That energy was convertible was not new; indeed, humans had applied this principle for centuries without knowing it. For example, they knew how to use the chemical energy contained in wood to warm themselves or cook their food; they knew how to use the chemical energy contained in water to grow their crops. They knew how to use the hydraulic energy of waterfalls to turn the grindstones of their mills, and they even knew how to use the kinetic energy of the wind to move their sailing ships to the other side of the world. People had done all these things for a long time.

However, they had not established the relationship linking all these phenomena. Similarly, they did not know that they could use fire for purposes other than to warm themselves and their food or melt their metals. They knew that fire could produce heat, but they had not yet realized that it could also cause movement, that it was the fire that made the lid dance on the top of the pot. It was not until the beginning of the nineteenth century that people finally realized that it was the steam generated by heating water on a fire that made the lid dance, and if steam could make a lid dance, it could also make pistons move and turbines turn, resulting in work. Fire had not only a heating power but also a motive power. Moreover, this power could be controlled, to a certain extent.

Once mankind realized that they could use the motive power of fire, they were no longer dependent only on their own strength or that of their beasts of burden to do work. They could use the motive power of fire, as they had used the motive power of the wind or water, with this difference: this power could be generated where they wished and when they wished. They needed only fuel and a match to make the world move. This awakening radically affected humanity, perhaps as much as the discovery of bronze during the Stone Age. The industrial revolution had begun, and modern science soon followed.

Fire transformed water into steam and steam produced motion. This motion could be used to push a piston, which could be used to make a wheel turn, which could start an engine, make a machine turn, and even propel a locomotive. The French physicist Sadi Carnot (1796–1832), who sought to determine how to maximize the power of fire to do work, imagined a steam engine whose efficiency would be optimal. He published his discoveries in 1824, in a work titled *Reflections on the motive-power of fire and the machines suitable to develop this power.*

Carnot's calculations led him to an unexpected conclusion. He noticed that while, in theory, all the hydraulic power of a waterfall or all the kinetic energy of the wind could be transformed into motive power (mechanical energy), it was not possible to transform all the heating power of fire (thermal energy) into motive power, even by building an ideal machine. When work was "extracted" from fire, there was an inevitable loss of energy, inherent in the nature of fire.

At first sight it would appear that the second law, which stipulates that there is an inevitable loss in the transformation of heat (thermal energy) to work (mechanical energy), has nothing to do with atoms and molecules. As we already said, until the beginning of the twentieth century many scientists still denied their existence. It was possible to interpret the observations made prior to that time without relying on the concept of atoms and molecules. Even in the observations and the conclusions of Carnot, molecules were not necessary. In fact, although his conclusions were correct, Carnot's own image of heat was wrong. It was based on a completely false and widely accepted concept, the *caloric.*

The idea of the caloric, much in fashion at that time, had been introduced by Lavoisier and developed by the physicist and chemist Joseph Black (1728–1799). This idea was adopted by many scientists, including Carnot. According to this theory, which survived for almost two centuries, heat was an invisible fluid without weight, the caloric, which penetrated every body. The higher the temperature of a body, the more caloric it possessed; and just as fluids always flow from a higher to a lower level, the caloric always passed from a hot body, with a high caloric, to a cold body, with a low caloric. It was convincing.

However, Ludwig Boltzmann (1844–1903), an Austrian physicist and mathematician, was not convinced. He rejected this image of heat. Boltzmann believed in molecules. He had adopted the atomic theory suggested by the Greeks and put forward decades before thanks to the work of the English chemist John Dalton (1766–1844) and the Irish physicist Robert Boyle (1627–1691). Boltzmann believed that heat was due to microscopic movements. This theory was based on several observations, in particular on an experiment that shows that snow melts more quickly when it is rubbed with another piece of snow. Boltzmann believed that these microscopic movements were caused by very small and numerous particles of matter constantly knocking each other in all directions: the molecules. Today, heat is conceived of as the kinetic energy of molecules. By assuming the existence of these particles of matter and their ceaseless motion, the second law, which, in fact, stipulates that one cannot use the power of fire without "paying a tax," no longer seemed to be a divine punishment, but a simple principle that could be explained logically. This explanation was given by Boltzmann and is clearly described by Peter Atkins in his great book about the second law.[1]

To understand the second law, imagine that thanks to the fire you have lit under a pot filled with water, the water heats. As water heats, water molecules move faster and faster, gaining more and more energy from the fire. At the boiling point (212°F), the water molecules are fast enough to all leave the pot and reach the air; in other words, it evaporates. As the water molecules reach the air, they are scattered in all directions.

In the air, these fast molecules have enough energy to push whatever stands in their way, to the point where they can even propel a train (a steam train!). However, to do this work molecules have to push in the same direction, the direction of the desired motion. (Just as a number of people must all push in the same direction to move a broken-down automobile.) In other words, the molecules that reach the air in the direction opposite to the desired motion are not part of the action. They can't do the job. Although they have also gained kinetic energy from the fire, this energy cannot be used and therefore is lost from the system.

Carnot discovered that manipulating fire to extract its motive power is like manipulating a raw material to extract a shape from it: some loss of material (sawdust, chips, crumbs, and so on) is inevitable. The wood chips do not disappear, but for the carpenter, they are almost unusable. In the same way, in the extraction of the motive power of fire, there is an inevitable loss; a loss of the energy of the molecules that do not contribute to the desired motion. This "lost"

Figure 7.1. When carving, some loss of material is inevitable. Though the chips do not disappear, they are no longer very useful.

energy does not disappear from the world but it becomes practically unusable—at least by the system.

Carnot's work became the basis for formulating the second law of matter, the second law of thermodynamics. This teaches us that even if energy is never lost, it can become unusable as a consequence of our manipulations. The energy contained in our natural resources (wood, oil, and so on), which we constantly extract, is dwindling and is becoming less and less exploitable. Furthermore, just as it would be unprofitable for the carpenter to collect the sawdust and glue it back together, today it is still unprofitable to us to gather this dispersed energy. To do so would require an effort that would not be cost effective. Though energy can never disappear, the amount of energy we can use today is decreasing, and this constant loss is at the origin of the energy crisis that threatens the world.

Boltzmann's molecular conception of the second law of matter went much further than Carnot's steam trains. It related to the movement of molecules in general. Boltzmann discovered that this movement was following the most fundamental laws of nature, the laws that govern the entire universe, the laws that should awe us by their very existence: the mathematical laws. Indeed, whereas many continued to view molecules as fictitious, Boltzmann, who firmly believed in them, introduced the probability laws of large numbers to describe their movement. He showed that the second law of matter was the logical consequence of the spontaneous movement of a population, the number of whose members was as large as the number of molecules.

Consider, for instance, a deck of cards. Shuffle them and look at them in a row. Do you see five cards of the same suit (a flush) lined up? I doubt it. It is quite rare. It is even rarer to see five consecutive cards of the same suit (a straight flush). Actually, it is quite improbable. This is why a straight flush has more value than a flush, because it is rather rare. And in poker, as in all card games, the rarer the hand, the higher its value. Now recall our experience with the boiling water and imagine that you are able to photograph the billions of molecules scattered in the air and note their direction at the moment when the picture was taken. What chance would you have of finding the molecules heading in the same direction? Practically none. You can repeat the experiment and photograph these molecules all day, for the rest of your life, but you would never obtain a photo where all the molecules or even part of them are arranged in neat rows, all going in the same direction. You would always see chaos or molecular disorder, because this is more probable. You are unlikely to find that molecules have spontaneously arranged themselves in order, because if, with regard to a few cards drawn randomly, an ordered arrangement is less probable than a disordered arrangement, with regard to a population as large as that of molecules, the chance of an ordered arrangement is practically zero.

In the molecular world, when energy is released by a chemical reaction, it is as if a "price" has been paid, and order can be achieved. In the molecular world, order can be generated, for example, via a reaction that lowers the number of molecules: consider the formation of water in figure 7.2. In this reaction two molecules of hydrogen, H_2, react with one molecule of oxygen, O_2. These three molecules all together give two molecules of water H_2O.

Such a reaction is said to generate order and cannot occur if no energy is released. It has to be emphasized here that not all the reactions that release energy make order. However a reaction that makes order *has* to release energy, otherwise this would mean that order was made "for free," and the second law teaches us that order costs energy.

On the other hand, chemical disorder can be made through decomposition, disintegration, dissolution, mixing, and so forth. For instance, when you put a piece of sugar in your coffee, the sugar dissolves and water molecules mix with the sugar molecules; molecular disorder is thus increased.

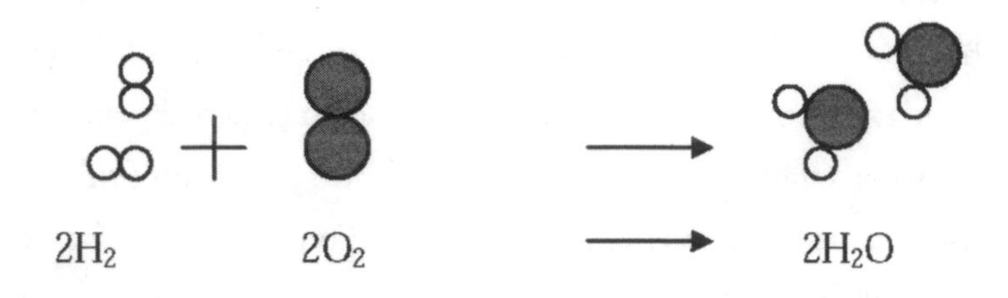

Figure 7.2. Two molecules of hydrogen, H_2, react with one molecule of oxygen, O_2. These three molecules altogether give two molecules of water, H_2O.

In fact, the second law can be seen as a truth that confronts us daily in all areas of our lives, namely that to generate order, you sacrifice energy. Order can't happen by itself. This is true for Nature and for us. If you are not convinced, just think about your desk or closet. If you do not invest the effort necessary to put your room in order, disorder will only increase. This second law reveals that even Nature, as majestic and powerful as it is, cannot escape such a banal and prosaic principle: to make order requires energy; disorder occurs by itself.

Our universe is like an immense network of molecules. I use the word *network* because, although these molecules appear like individual molecular clusters (that is, human beings, animals, plants, solids, liquids, and gas), these clusters are not isolated from each other. There is no precise demarcation or border between the components of our universe. For example, there are no fences between the molecules of a tree and those of the air that touches it. If there were, what would they be made of? If we keep in mind that everything moves, we can guess that, at least in the air around the tree, we would find molecules coming from the tree. There is no fence between the molecules of the page that you have in your hand and those of your fingers that hold it; in fact, your molecules are so numerous that, without your being aware of it, you are losing thousands of them all the time. They remain stuck to all that you touch, including this page; the proof is that your fingerprints are detectable there.

In the case of solids, whose molecules are slow and bound tightly, exchange of molecules with the outside is limited. For example, there is very little exchange between the molecules of a plate and those of the table on which it sits. On the other hand, the molecules of bodies that are less dense, more mobile, and touch each other. When we go out for some fresh air, for example, the molecules of air penetrate our skin even if we can only breathe through our nose and mouth. In the same way, when we take a shower, water penetrates our body more deeply than we might imagine. In other words, we can consider that all the components of our universe that are in direct contact with each other are exchanging molecules, if only very few, and can thus picture our universe as a three-dimensional network, no part of which is isolated from the whole.

A chemical reaction would be like a disturbance at a very specific spot in this molecular soup, like a tiny stir. If such a stir releases energy, local order might appear. Nevertheless when we look at the whole picture and consider our entire universe, according to the first law, energy cannot disappear from it. In other words, our universe as a whole cannot lose energy, and therefore even if energy is released to the outside by a particular reaction and order is achieved locally, the overall molecular chaos (or molecular disorder) of the world still increases at each reaction, because the energy of the world is constant and no energy can be released. There is no "outside." The second law stipulates that any reaction, any stirring in this molecular soup that is our world, increases global molecular disorder.

In a certain way, a chemical reaction may be seen as the shuffling of cards. After they are shuffled, cards are usually less ordered than before, and the more they are shuffled, the more their disorder increases. This is because it is more probable that they become more disordered than ordered. And it is even more probable if you don't have fifty-two cards, but billions and billions and billions of them.

Indeed when you look at a glass of Coke, you know that it is full of sugar. However, you do not expect to see a piece of sugar emerging from your glass. Why? Because of the second law. Because the probability that enough of the sugar molecules that are mixed with innumerable water molecules will organize and move in the same direction to find themselves close enough to each other to coalesce into a piece of sugar, is . . . zero . . . unless they arrange a rendezvous by telephone. And telephone calls are not free.

If, for us, *disorder* is a vague and unspecified term, for chemists molecular disorder is represented by a precise and measurable physical parameter called *entropy*. The higher the entropy of a system, the more disordered it is. This physical parameter was first introduced in 1865 by the German physicist Rudolf Clausius (1822–1888) and corresponded to a measurable entity: the quantity of heat absorbed or released by a system divided by its absolute temperature.

Not until Boltzmann's work was it clear that this physical parameter also reflected the extent of the molecular disorder of a system. In a formula that bears his name, Boltzmann established the algebraic nature of the relationship between entropy and molecular disorder. Molecular disorder thus became measurable. Despite his genius, Boltzmann's ideas were disputed for a long time by his colleagues. Lonely, sick, and depressed, he hung himself in 1906 at the age of sixty-two, shortly before his formula became one of the pillars of modern science.

Incontestably, human beings are the result of a fabulous process of spontaneous order and organization. Given basic nutrition, our organism is able to manufacture, in cooperation with the molecules of the opposite sex, sometimes even without our knowledge or approval, an extraordinarily sophisticated product that one day will walk, speak, and think. But that does not mean that our existence can avoid the verdict of the second law. Our body unceasingly emits gases, liquids, and other waste. It also diffuses heat because its temperature (98.6°F) is generally higher than the outside temperature, at least in Jerusalem, but if the temperature outside is high, we tend to "evaporate" water by sweating. In all these processes energy is lost and the molecular disorder in the portion of the network we occupy increases and more than compensates for the molecular order that we represent. This is how our presence also contributes to the increase of the molecular disorder in our universe.

In summary, the second law is nature's lugubrious message to us: interactions between molecules, whatever form they take, inevitably increase molecular disorder. With each chemical reaction, the order among the molecules that compose our universe decreases, because it is much more probable. Thus, after a reaction has occurred, the molecules involved in the reaction have intermingled a little more, and the molecular network at that place has become slightly more uniform. With each reaction, the universe loses some of its shape, like a building that slowly disintegrates. It becomes more homogeneous, more uniform, simply because by doing so, it reaches a more probable state—our immense universe is unexpectedly governed by mathematical laws, the laws of statistics and probability.

However, if the second law of thermodynamics foresees that the disorder of our universe can only increase, it does not exclude the appearance of a phenomenon as ordered and thus as improbable as life. In this world, dominated by the laws of probability and by statistics, there is room for the unforeseeable and the unexpected.

Nor is that all. Behind these messages there is yet another. With each chemical reaction, entropy increases and living or inert matter evolves. With each chemical reaction, molecules leave a past they cannot return to. The entropy of the universe will never again be what it was yesterday. When you dissolved your sugar in your coffee this morning, you contributed to the increase of the entropy of the universe, and, without noticing it, you continued to do so all day.

If data on the entropy state of the universe could be recorded at the precise time of historical events, it would be easy, without even knowing what they were, to order these events chronologically. The one whose entropy is lowest would be the oldest, and the one with the highest entropy would be the most recent. The direction of time is always from the lower to higher entropy. Matter is said to carry in itself "the arrow of time," an arrow that shows the direction of time, an arrow that would have been launched at the very moment of the birth of our universe, and which, since then, nothing and no one can stop.

In short, let us say that, since its birth, matter has had, like us, a sense of time. The opposite would be surprising because, after all, what are we other than matter that moves and speaks? With its portent of catastrophe, this law has caused much ink to be spilled, not only by scientists but also by philosophers. If matter, according to its laws, moves toward an irrevocable destiny, and we can never escape from it, what power do we have? Where does the mind come in?

8

THE THIRD LAW

Perfect Order Would Be Cold!

It is very difficult to speak about entropy in absolute terms. We usually speak about the difference in entropy before and after a reaction. The alternative would be to refer to the birth of matter and of entropy, that is, to the beginning of the universe. The third law helps us fix a more practical reference point for entropy than its remote birth.

The third law determines the conditions under which an isolated system would have an entropy equal to zero. It stipulates that a system with zero entropy corresponds to a crystal made up of identical atoms, at absolute zero (−459.67°F), a temperature that can never actually be reached. At this temperature there is neither motion nor space between the atoms. Identical and motionless atoms are arranged side by side, in perfect rows. For chemists, this motionless arrangement corresponds to perfect order. Actually there exist no such systems with zero entropy because, as we said, absolute zero cannot be reached. Above this temperature, movement is triggered and disorder begins.

The opposite of perfect order, when entropy equals zero, would be maximum disorder or total chaos, a state where nothing can happen anymore. In such a state no chemical reaction could occur because the entropy of the universe would have reached its maximum value and could no longer increase. At this stage, the texture and the color of the whole universe would be perfectly homogeneous, as if the whole universe had been blended in a mixer.

Note that the concept of order and disorder is very subjective. It is possible to imagine, however, as the British writer J. E. Lipman says, that perfect order corresponds precisely to maximum entropy, that is, a perfectly homogeneous universe.

9

LIFE IS NOT RESTFUL

As explained earlier, a chemical reaction is a disturbance at a specific location in the infinite molecular network that is our universe. At this specific location the network is modified; its composition is affected, and sometimes also its temperature, texture, color, odor, and so on. At the same time, this disturbance causes an increase in the global molecular disorder. But a chemical reaction has an end. The disorder that reacting molecules can produce is limited. When a reaction is finished, it is because at this specific location, under the newly created conditions, molecular disorder can no longer increase. It has reached its maximum. This corresponds to the most probable state that could be created under the given conditions. This relative "at rest" state is called *chemical equilibrium*. I will call it here just *equilibrium*. I have referred to it as a *relative* state because the molecules remain agitated by Brownian motion, but the portion of the network that they occupy does not change its composition, its texture, its temperature, its color, and so on. The system has reached its most probable molecular configuration.

For example, when sugar is put in a cup of hot coffee and stirred, the sugar melts and the coffee cools down. These two phenomena, the sugar mixing with the coffee and the hot mixing with the cold, cause an increase in the disorder (now we can say *entropy*) of the universe. Disorder increases until the concentration of the sugar is the same everywhere in the cup, and until the coffee reaches room temperature. When this is achieved, entropy can no longer increase, not without an external perturbation. The chemical system is at equilibrium.*

Now consider living organisms. From the simplest to the most complex, living organisms consist of cells, sometimes one cell like a protozoan, sometimes several billion like our body. These cells are made of molecules (of what else could they be made?). The human body contains approximately 50,000 billion cells of 220 different types corresponding to the various types of tissues that form our bodies. These cells are centers of intense activity. The heart cells, for example, act as small "engines" that use energy drawn from the reaction between carbon and oxygen to make the heart contract and pump blood through the arteries, the capillaries, and the veins. This mechanism operates through innumerable chemical reactions involving the molecules of the cells. Moreover, these cells are constantly being destroyed and reproduced as part of their chemical activity. To maintain the intense activity of its cells, a living organism is in constant contact with its surroundings. It draws from them the energy the cells need to do their job and it discharges waste and dead cells. The survival of an organism depends on these exchanges, which occur only through chemical reactions. This endless activity prevents an organism from reaching chemical equilibrium.

In other words, as opposed to inert matter, a living organism is not in equilibrium with its surroundings. For example, if our body were in equilibrium with its environment, its temperature would continually vary in order to be equal to the surrounding temperature. A living organism is not in equilibrium; it never rests. It is like

*This is not totally true, because water evaporates slowly from the cup, but for our purposes here we can ignore this slow process.

a factory that never stops. Day and night, 24/7, its molecules are busy forming and breaking chemical bonds. Your molecules are constantly reacting even when you are not totally aware of it; even in your sleep you breathe, you sweat, you move, and so on. The major difference between inert matter and living matter is that a living organism is never at equilibrium . . . for as long as it lives. In fact, for us, as for any other living system, chemical equilibrium is synonymous with death.

Imagine that you, lying on your bed, decide to deprive your body of all energy food, light, heat, and so on. In other words you decide to deprive your body of the energy it needs to stay far from equilibrium. What would your molecules do? Where would they go? Without a supply of energy, you would quickly become sick because of faulty reactions. Your cells would decay, their structure would collapse slowly, your body would break down and become absorbed by the environment. Your molecules would disperse, creating the maximum disorder possible according to the second law of thermodynamics. At the end of this process, equilibrium would be reached. Your skeleton would remain, naked, and its temperature would be that of the room.

We can, therefore, deliberately cease our molecular activity by depriving our organism of the energy necessary to maintain it far from equilibrium. Suspending our consumption of energy would modify our molecular activity and would make us slowly glide toward equilibrium.

We know all too well that our death is inevitable. But we usually do not know that this death sentence was signed at birth by the second law of thermodynamics. It teaches us that equilibrium is the most probable state a system can reach, not life. Life is not a probable state for a chemical system—death is. Indeed, of all that can happen to you in life, isn't death the most probable?

It is by maintaining exchanges with our surroundings that, to some extent, we are able to postpone this inevitable end. While inert matter in reaction moves toward equilibrium, living matter "avoids" it as long as possible. This is the fundamental difference between inert matter and living matter. The general direction of their molecules is diametrically opposite. During each chemical reaction, inert matter is heading toward equilibrium, toward the most probable state, that is, toward increasing molecular chaos, while living matter—by maintaining exchanges with its surroundings—heads in the opposite direction, toward the improbable, toward organization and life that breeds life.

Going from the most primitive to the most complex organisms, we can see that the strategies used to maintain exchanges with the surroundings and thus avoid equilibrium seem to become more sophisticated. Plants, for example, limited in their movements, are content to absorb those substances found in their immediate neighborhood to maintain their molecular activity. Animals have the ability to move in order to get what their organisms need to maintain themselves far from equilibrium. Human beings can speak and express their needs.

The origin of life, the evolution of living matter from inert matter mentioned at the beginning of this book, could be the result of an extraordinary journey of molecules away from equilibrium. In this context, metabolism and even reproduction, the two most specific phenomena of the living world, could be seen as highly sophisticated strategies used by molecular systems to maintain themselves far from equilibrium.

In summary, two possibilities seem to dwell in each of the molecules of living matter during its lifetime: the death instinct, which is engraved on all molecules, corresponds to their attraction to equilibrium, the most probable state; and the life instinct, reflected in the cooperation of molecules in maintaining the integrity of their molecular edifice. This is expressed by their attraction toward improbable states, to uncertainty (not to say adventure), where ever-renewing life resides. Indeed, if our own life is ephemeral, life itself seems as immortal as a rock.

It is often said that speech is what most differentiates humans from all other animals. However, we seem to have something more, which no other living thing possesses: choice. Undeniably, we may choose to stay alive or put an end to our lives just by lying on a bed and waiting for death. At every moment throughout our lifetime, we choose either to succumb to the attraction of our molecules to equilibrium or to continue trying to avoid it. At every moment we choose between life and death.

PART TWO

The Secrets beyond Matter

10

BIOLOGICAL INFORMATION

Do You Hear Molecules Talk?

In the enormous molecular network that is our universe, living organisms, from the smallest bacteria to the human, are made up of cells. These cells, made of molecules, are the sites of intense chemical activity generated by the molecules. During a lifetime, this activity prevents a living organism from reaching equilibrium with its surroundings. Without this ability for self-maintenance far from equilibrium, all the molecules of a living system would move directly toward equilibrium and it would die. Today, as far as science is concerned, this ability is a miracle, the miracle of life. It is the privilege of the living world. Throughout its lifetime, the molecules of a living being receive a deferment from being dispersed. During its lifetime, they ensure its integrity.

We could suppose that, as there are four natural forces that prevent our universe from breaking down, there might be a fifth that keeps the molecular assemblies, the living systems, from falling apart by preventing them from reaching equilibrium and disintegrating. We could call this force the *vital force,* that which, during our lifetime, holds the cells of our organism together. However, like vital energy, this vital force is not part of the scientific vocabulary. Even supposing that it exists, it cannot yet be defined by a formula, much less measured. It can be manipulated to a certain degree—it can be reproduced or destroyed—but its source is unknown and no one can create it. Moreover, for the moment, science continues to develop quite adequately without it.

To ensure its integrity and functioning, a living organism does business with its environment. This trade can operate only through chemical reactions. All that we swallow, that we breathe, that we touch, that our ears, our nostrils, and our eyes perceive, in short all that our five senses may grasp, leads to chemical reactions. Even the least of our movements, the least of our sighs, the least of our thrills, is the result of an endless number of intricate chemical reactions. In fact, like any living organism, ours is preoccupied with only one thing: the formation and the rupture of chemical bonds, that is, the production of chemical reactions. These contacts with the environment are vital, but they can also make us sick and even kill us. An infectious disease, bacterial or viral, always corresponds to an "enemy invasion," penetration by intruders who take control of part of the activity of our cells by modifying the course of some of their chemical reactions. The medicine we take is a judicious chemical whose purpose is to repair the damaged portion of the network by forcing the undesirable molecules to follow other trajectories and finally be transformed or leave the organism.

In a healthy body, the network of cells is intact and the activity of the cells is well coordinated, like the activity of the workers of a model factory. This coordination is ensured by the transmission of information. This information—which could be seen as the instructions that the cells must follow in order to stay together, grow, multiply, and fulfill their functions—is transmitted by molecules. These instructions are written in a "language" that was discovered only a few decades ago, one whose decipherment has hardly begun, a language spoken by molecules.

Biologists have concluded that the molecules of living organisms have the capacity to store and transmit information by means of their reactions. This constant communication between biological molecules allows the correct functioning of the cells. Today, only a small number of these molecular messages can be deciphered, because the language used by the molecules is so subtle and so sophisticated that the decoding of the smallest bit of information requires several years of research in this relatively new field called *molecular biology*. Biologists think that information is stored in the structure of the molecules of living matter. Compared to molecules of inert matter, the structure of the molecules of living matter is very complex; they generally contain a very large number of atoms: several thousand and even several million. It is usually accepted that the more complex the structure of a molecule is, the more important the information it contains.

DNA, for example, is a molecule that contains millions of atoms. DNA is found in all the cells of living organisms, from the simplest to the most complex. We know today that DNA is the central and largest data bank of any organism. According to biologists, it contains in its structure all the information necessary to recognize and duplicate each of our cells. Our DNA "knows" all about us: our size, the color of our eyes, of our hair, and of our skin. Our DNA knows our constitution and its resistance. It also knows all about our ancestors. It contains all our genetic past, from generation to generation.

In fact, the term *DNA* does not refer to a specific molecule but to a group of molecules. Imagine a group of necklaces all made of four kinds of pearls. From one living species to another, it is only the length of these necklaces and the order of the pearls that varies. For example, it has been recently discovered that only 7 percent of our

DNA is different from the DNA of a macaque. Between humans, DNA is 99.9 percent identical: this means that only 0.1 percent of the DNA of any two human beings is different. The difference between your DNA and that of your parents is even smaller.

Nevertheless, each human being's DNA is unique. Throughout human history, there has never been another being like you, with exactly the same DNA, unless you have an identical twin. And there will never be another being like you. You are unique; and as long as the cloning of human beings remains a fiction, you will remain unique until the end of time. Thus, your DNA distinguishes you from all the other beings on our planet. Don't worry, then, if sometimes you feel different from the rest of the world; you are. In each of your 50,000 billion cells is something that no one on earth possesses but you, and that no one can take from you. We may be merely made of nearly empty atoms, but nevertheless science confirms that each of us is unique.

Biologists believe that information passes from one molecule to another by means of reactions in which molecules "touch" one another. These interactions can be conceived as the meeting of two molecules with complex structures: one that has the shape of a "lock" and the other the shape of a "key." Only when the key comes into contact with the right lock does a reaction take place and information is transmitted. In other words, the information carried by one molecule is transmitted to another only when this molecule has been "identified" by its structure. This arrangement ensures the selectivity necessary to maintain order between constantly reacting molecules. Without this selectivity, we might grow nails under our tongues or teeth at the tips of our fingers.

Compared to living matter, inert matter has been regarded as "mute," that is, unable to transmit information. The Belgian chemist Ilya Prigogine (1917–2003) came along to unplug our ears, as we shall see in the next chapter.

11

FAR FROM THE CROWD

As we have seen, inert matter can be found in two states: "at rest," meaning at equilibrium (even if temporary), and "in motion," meaning reacting. Living matter can be found in one state only: in motion. It is never at rest. It is constantly reacting. Moreover, its motion doesn't lead it to equilibrium: it seems to "avoid" it, because for living matter equilibrium means death. On the other hand, as soon as inert matter is in motion, it tends toward equilibrium, toward the most probable state. It is as if inert matter were at the edge of an abyss, with equilibrium at the bottom: the slightest motion will make it "fall into the abyss."

But this concept is incomplete, as was shown only a few decades ago. Researchers discovered that inert matter in motion did not always move directly toward equilibrium; it sometimes made a detour. It was as though, under certain conditions, it left its usual path leading directly to equilibrium and went off on a winding trajectory. Imagine a stone that, dropped into an abyss, does not fall straight down but rather falls like a slalom skier descending through a snowy forest. Researchers discovered that certain chemical systems in reaction could, by these detours, delay their fate and temporarily maintain themselves far from equilibrium. There they exhibited some very strange behavior.

One of the first descriptions of this type of system was given by the Russian biophysicist Boris Pavlovich Belousov (1893–1970) in the 1950s. Belousov was looking for a model for the Krebs cycle, one of the metabolic processes that lead to energy production. He mixed the following chemical compounds: a colorless solution of potassium bromate, a colorless solution of citric acid, and a yellow solution of cerium sulfate. The resulting mixture, of course, was yellow. As an informed scientist, Belousov expected that a reaction would occur, at the end of which cerium atoms would gain an extra electron and their color would disappear. In other words, as the reaction progressed, he expected to see the yellow color disappear and the solution to become colorless.

The solution indeed became colorless but then, amazingly, it became yellow again! And then colorless! And then yellow! It was as if the solution oscillated a few times between the two colors. Imagine a glass of water turning into lemonade, the lemonade turning back into water, and then back into lemonade, and then back into water, and so on, spontaneously, before your eyes. The changes continued until, at some point, the reaction stopped, and one of the colors prevailed. The effect was quite stunning.

Belousov submitted these results for publication. However, they were so weird that they were immediately rejected. No one believed him. Much effort was necessary to overcome a virulent skepticism, of which traces still remained in 1972. For many, the phenomenon was unbelievable because it seemed to violate the second law of matter. The phenomenon he described, during which molecules seem to work in unison, could only be the result of a spontaneous molecular organization in space and in time (since the phenomenon was periodic). However, according to the second law of matter, under the conditions of the experiment, such a spontaneous organization of reacting molecules—which would have meant that the entropy of the system decreased—was impossible. However, it was shown later that other systems exhibited similar behavior.

Ilya Prigogine was among the first to explore this strange behavior of matter. He succeeded in providing a mathematical model that explained how a chaotic chemical system could, in certain cases, spontaneously remain far from equilibrium and generate molecular organization without violating the second law of matter. He called these systems *dissipative structures*. He showed that although entropy dropped as these structures formed, this was a local drop; at no time did the overall entropy of the entire system decrease. Therefore the appearance of dissipative structures and their organization did not violate the laws of thermodynamics.

Prigogine explained that dissipative structures were specific systems possessing particular characteristics. He pointed out that some of the characteristics of these specific chemical systems and their dynamics were reminiscent of those of living systems. Like living systems, they manifested organization and were "animated" by a complex network of many different chemical chain reactions triggering each other (for example, compound A reacts to create compound B, B creates C, C creates A, and so on). This complex chemical activity maintained those systems far from equilibrium.

Like living systems, they were also consuming energy. Indeed, to stay organized, these systems had to be constantly provided with chemical energy and, like living systems, they were temporary: if no energy was supplied, these structures ceased to appear, and the phenomenon "died" as the system reached equilibrium. Prigogine went even further, declaring that the element of unpredictability characteristic of these structures was related to the notion of creativity in living organisms (*creativity* is defined here as "the capacity to produce unique phenomena").

Prigogine concluded that life had its origin in these kinds of systems. In a molecular network condemned to a slow journey toward an ever-increasing molecular disorder, dissipative structures had emerged from outlier molecules, against all odds. Spontaneously maintained far enough from equilibrium, thanks to ongoing chemical activity, these chemical systems would have evolved slowly into more organized and sophisticated structures until they became living organisms able to perpetuate life. Nor is this all. He writes:

> Such a degree of order stemming from the activity of billions of molecules seems incredible, and indeed, if chemical clocks had not been observed, no one would believe that such a process is possible. To change the color all at once, molecules must have a way to "communicate." The system has to act as a whole. We will return repeatedly to this key word, communication, which is of obvious importance in so many fields, from chemistry to neurophysiology. Dissipative structures introduce probably one of the simplest physical mechanisms of communication.[1]

And few pages later:

> One of the most interesting aspects of dissipative structures is their coherence. The system behaves like a whole, as if it were the site of long-range forces. In spite of the fact that interactions among molecules do not exceed a range of some 10^{-8} cm the system is structured as though each molecule were "informed" about the overall state of the system.[2]

Imagine a swarm of flies in chaotic movement suddenly organizing itself spontaneously, like the dancers of the Red Army Dance Ensemble in a final scene, and you have an example of the coherence to which Prigogine refers. For Prigogine, the formation of dissipative structures was the result of communication between molecules. In order for these molecules in random motion to spontaneously react in such a coordinated way, they had to communicate, just as the members of the Red Army Dance Ensemble do by hearing and speaking. He concluded that the state of non-equilibrium contained the potential for communication. Just as molecules of living matter communicate in order to organize, the molecules of inert matter, far from equilibrium, might also communicate. In this view, moving away from equilibrium was for the molecules like people moving away from the hubbub of a crowd and suddenly being able to hear and talk.

For the first time in the history of modern science, communication among molecules of inert matter was mentioned. However, although Prigogine was awarded the 1977 Nobel Prize for chemistry for this work, his opinions were never very popular among scientists. Neither his ideas about the origin of life (many believe today that the first step in the transition from inert matter to living matter is the formation of self-reproducing molecules and not the formation of dissipative structures) nor his remark on communication between molecules retained the general attention of the academic world. How could molecules of inert matter communicate?

As mentioned earlier, biologists believe that molecules of living matter recognize each other by their shape, and thus communicate "by touch." When a molecule meets another molecule whose structure corresponds to its own, like a key to a lock, a reaction occurs and information is transmitted. This means of communication is inappropriate for the molecules of inert matter such as those involved in dissipative structures: they are too small (with few atoms, compared to the tens of thousands of atoms in biological molecules), and therefore their structures are too simple and unsophisticated. They cannot generate the necessary selectivity for the kind of communication described by biologists. Without selectivity, molecules would easily lose their way.

We communicate by the sense of touch, sometimes even by the sense of smell, like many animals, but waves such as sound waves and electromagnetic waves appear to be a very broad means of communication.

We should keep in mind that electromagnetic waves are at the basis of all the new means of communication we have invented—the telephone, radio, television, satellite, Internet, and so on. Could molecules also communicate through some kind of waves? This brings us to the discoveries that shook up science in the twentieth century and totally blurred our image of matter.

12

THE QUANTUM REVOLUTION

Matter's Double Nature

Thanks to the genius of Sir Isaac Newton (1643–1727), from the seventeenth century until the beginning of the twentieth century we thought we knew all about the movement of solids. From dust to stars, this motion was described with great precision by Newton's laws. What we still ignored was the nature of the matter from which the stars and dust were made. We were far from suspecting that its discovery would completely change our view of the universe.

At the beginning of the twentieth century, just before Boltzmann committed suicide, Einstein finally ended the argument over atoms in an article on Brownian motion published in 1905. The existence of atoms and their molecules was undeniable, and the atomic theory was gaining acceptance. Scientists began studying atoms more closely.

In 1904, the British physicist Joseph J. Thomson (1855–1940) discovered that matter contained negative charges. Since atoms were electrically neutral, he hypothesized that the atom was a positive sphere that was stuffed with these negative charges (the electrons) like a raisin muffin. Thus the atom's electrical neutrality was maintained. In 1911, his student from New Zealand, Ernest Rutherford

(1871–1937), showed that there was indeed a positive sphere in the atom, but this was very dense and contained most of the atom's mass. Electrons were not stuck to it but were far away. An atom no longer resembled a raisin muffin, but a fruit with a kernel at the center. On the other side of the ocean, the American Robert A. Millikan (1868–1953) combined these results with his own and succeeded in 1909 in calculating the mass of an electron. Later came the discovery of protons and neutrons by the British physicist James Chadwick (1891–1974) in 1932.

In short, by the second decade of the twentieth century these experiments had enabled scientists to create the atomic model briefly depicted in the first part of this book: an atom contains a positive, very dense nucleus. This nucleus is composed not only of positively charged protons, but also of neutral charges, the neutrons. Surrounding these, very light negative charges, electrons, move at great speed and in a tremendous void compared to their size.

This was the only possible model consistent with the results of the observations made and confirmed by researchers on both sides of the Atlantic Ocean. However, scientists knew that something was wrong with this model. It could not work. According to the model, electrons revolved in orbits around the atom's nucleus, like planets around the sun. And, according to both Newton's laws of motion and the Scottish physicist James Clerk Maxwell's laws of electromagnetism (1831–1879), the negatively-charged electrons revolving around the nucleus should have lost energy and then crashed very quickly into the nucleus, attracted by its positive charge. Obviously, this was not happening. Electrons remained in their orbits and did not crash into the nucleus. Something was wrong. The model was incorrect, but nobody had a better one.

The embarrassment of the scientists was enormous. The laws of physics that were considered infallible no longer operated. They could not explain why an electron doesn't fall into the nucleus like a fly into a pot of honey. Newton's equations, which had turned out to be as valid for a grain of sand as for the planet Saturn, did not apply here. For the first time in more than two hundred years, Newton's laws had failed. These concerns, together with other unexplainable experimental results, forced scientists to reconsider what they had believed to be the unshakable foundation of physics. This reconsideration gave birth to the quantum revolution.

After having ruled the roost for more than two centuries, successfully describing the motion of all bodies in the universe, from a grain of dust to galaxies, Newton's laws were superseded in less than two decades by the new laws of quantum mechanics. Until then, Newton's laws had done their job. They were not wrong but they were incomplete. They did not apply at the atomic scale. Matter that we know, and of which we are made, is indeed a collection of small entities—electrons, protons, and neutrons. But these entities could not be conceived as tiny solids to which Newton's laws applied. Great imagination was required to find new equations that would explain how the electrons, which revolve at tremendous velocities, do not lose energy and end their frantic run in a tragic but unavoidable fall into the nucleus.

At the beginning of the twentieth century, the work of the physicists Max Planck (1858–1947) and Albert Einstein (1879–1955) established that light had a double nature: in certain experiments it behaved as if it was a wave—that propagates and may be refracted by a prism—and in others, it behaved as if it was composed of myriad strange, small projectiles, invisible and without mass, the photons. These photons can be imagined as the drops of a beating rain that started pouring billions of years ago, when light was born together with the universe, and which have been falling on us since then, at the tremendous speed of 186,282 miles per second!

Inspired by these results, in 1924 the French physicist Louis de Broglie (1892–1987) presented the astonishing hypothesis that matter also had a double nature, a wave nature and a corporal nature. Like light, matter could sometimes behave as if it were waves and sometimes could behave as if it were made of small beads. De Broglie proposed the equation that made it possible to establish the link between these two behaviors. That equation suggested that all moving bodies, whether as fast and light as the electron or as slow and heavy as our own body, had a wavelength. This wavelength depended on the mass and the velocity of the moving body. In other words, any body, whatever it is, from dust to stars, could be regarded at the same time as both an aggregate of tiny beads and as a wave phenomenon, that is, something that has a wavelength. This duality had not been imagined previously because, according to the equation, the wavelengths associated with macroscopic objects were too short to be detected with the equipment available.

Nevertheless, de Broglie's equation stipulated that when you hold a stone in your hands, although you undoubtedly feel that you are touching a solid body, you are, in fact, also touching something that behaves like waves, even if we still can't detect it yet with our modern technology. This revolutionary description of matter aroused both interest and incredulity among scientists. As extravagant as it seemed, in 1927, three years after de Broglie presented his extraordinary hypothesis, it was confirmed: electrons had a double nature.

In an experiment set up to study electrons scattered from a metal surface, electrons did not behave like particles. They behaved like waves—they produced interference, one of the principal manifestations of the world of waves, unknown in the mechanics of solid bodies. According to the mechanics of solid bodies, if, for instance, two balls of modeling clay are thrown at each other, they will stick together and form a new body. It would be absurd to think that these balls could cancel each other out. Waves can. Darkness can be generated with light. Under specific conditions, if rays of light having wavelengths are shone through two slits made in a thin wall and the patterns that the rays make on a screen behind the wall are observed, there will be areas of strong light, as if the rays are sticking together (like the two balls of clay), and the crests of their waves are adding up. This is known as *coherence* or *constructive interference.*

However, there will also be areas where the light is dimmed or has disappeared, as if the waves neutralize each other, the trough of one wave cancelling out the crest of the other. This is known as *destructive interference.* In other words, adding waves together can destroy them. Interference is a wave phenomenon. Light can cancel light. Sound can cancel sound. This effect is often exploited to neutralize radio waves or to limit background noise by transmitting waves of the same frequency in order to cause destructive interference.

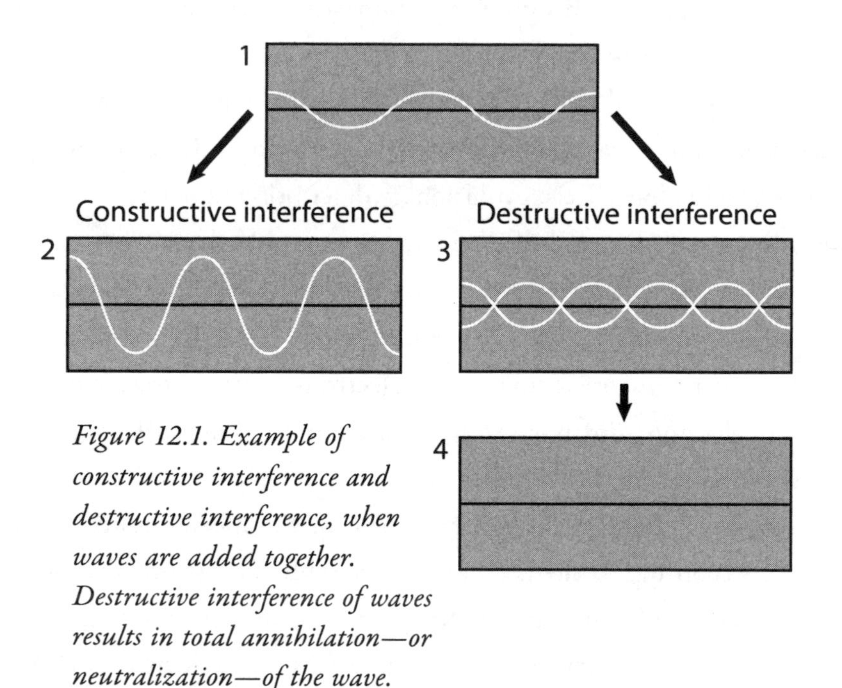

Figure 12.1. Example of constructive interference and destructive interference, when waves are added together. Destructive interference of waves results in total annihilation—or neutralization—of the wave.

Interference phenomena were observed accidentally by the Americans Clinton Davisson (1881–1958) and Lester Halgert Germer (1886–1971), when they were trying to calculate the energies of electrons scattered from a metallic surface. Electrons from a heated filament were accelerated and allowed to strike a metal surface that could be rotated. It was a great surprise to them to find that in this experiment electrons behaved like interfering waves, not like tiny solids. The appearance of an interference pattern was incontrovertible proof of the double nature of the electrons. At the same time, the physicist George Paget Thomson (1892–1971), Joseph J. Thomson's son, reached the same conclusions through different experiments. According to these experiments, electrons that had thus far behaved like solids with a definite mass and charge were suddenly acting like waves. The wavelengths associated with these electrons corresponded exactly to the wavelengths predicted by de Broglie's equation. Much later, similar experiments on accelerated protons and neutrons were carried out, and again de Broglie's hypothesis was confirmed. Researchers had to conclude that de Broglie was right; that, like light, matter was an enigmatic substance, something that can both behave as if it is made of tiny beads and sometimes produce interference as waves do.

So quantum physics was born, founded on this amazing hypothesis, which soon became an indisputable fact. The new field developed very quickly and was crowned with success. But it required a juggling act: on the basis of de Broglie's hypothesis, equations taken from the classical physics of solids (Newtonian physics) had to be combined with equations taken from the physics of waves. Thanks to the major contribution of the Austrian physicist Erwin Schröndinger (1887–1961), this maneuver was incontestably successful. The mixed equations, the most prominent of which is Schröndinger's equation, worked wonderfully. Experiments repeatedly confirmed their precision. This new theory provided spectacular results in multiple fields and opened new areas of research, one of which gave rise to the computer.

However, this mathematical manipulation had generated a monster, something that was neither really a wave nor a solid body. The bricks of our universe and of our own body, which had always had concrete and tangible form, had become a mathematical equation: the wave function. It was as if we were told that the coffee you drank this morning or whatever is around you right now is not what you believed it to be, but corresponded to $\Psi = \sum a_n \varphi_n$. At the atomic level, matter could not be described other than by this mathematical abstraction. No image was appropriate. There has not been much progress since then.

The inside of an atom can be imagined as an immense skyscraper with a huge number of floors, each corresponding to a different energy level, and with a basement inhabited by the nucleus. Electrons dwell on these various floors, according to their energy. The higher the energy of an electron, the farther from the nucleus it dwells. Since we have already compared energy with wealth, let's continue here and say that the richer an electron is, the more it can free itself from the hold of the nucleus. Just as money does not circulate as a fluid but as banknotes and coins, energy inside an atom does not circulate like a fluid but in discrete quantities, the quanta.

According to this image, a chemical reaction would be comparable to a financial transaction in which electrons lose or gain quanta of energy. When an electron is enriched by a reaction, it goes to the upper floors, and when it gets poorer, that is, when it loses energy, it descends to the lower floors, toward the nucleus. However, an electron can never "go bust" and lose all its energy (otherwise it would fall into the nucleus). On the other hand, if an electron gains enough energy, through a chemical reaction, it can overcome the attraction of the positive nucleus, that is, leave "its" atom and approach others. In other words, when its pockets are really full, an electron can "afford a trip."

Rather than using an elevator, the electrons pass from one floor to another by "jumping." These jumps are called *quantum leaps*. According to the theory, an electron can never be found *between* two energy levels; this means that during a quantum leap an electron disappears from one energy level and reappears on another. What happens to it in between floors? A mystery . . .

This very simplified view of an atom, roughly described here, is frequently used by physicists and chemists to explain many phenomena and make many experimental predictions, but it is obviously far from being exact, since the electrons are not actually small, solid bodies like those we're familiar with. In fact, we don't know precisely

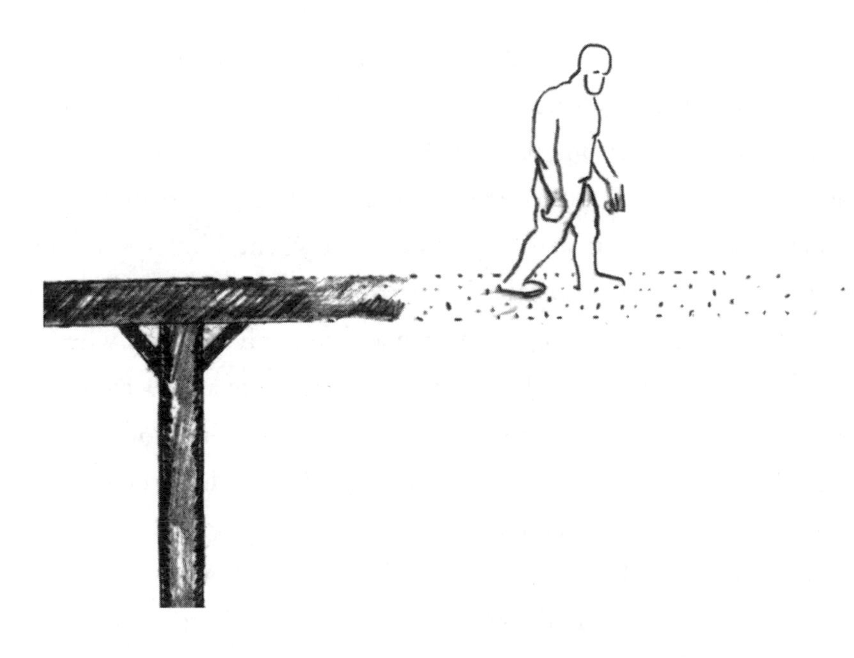

Figure 12.2. Leaping into the unknown

where an electron is located within its atom or what it looks like, nor have we any way at present to find out. However, electrons can be expelled and observed outside an atom. Have no illusions, though: the picture becomes no clearer. On the contrary, the plot thickens.

As mentioned above, electrons expelled from their atoms can generate interference patterns like the ones observed in the Davisson and Germer experiment: when a stream of electrons pass through a thin wall with two tiny slits, an interference pattern is observed behind the thin wall. In other words, the stream of electrons behaves as a ray of coherent monochromatic light. But what is mind-blowing is that when you close one slit and leave only one slit open, electrons behave as if they were particles: they pass through the only open slit and hit the screen behind like tiny beads.

With the two slits open, electrons behave like waves; with one slit open, like beads or bullets fired from a gun! The amazing thing here is that WE are the ones who decide to open a slit or not. In other words WE are the ones who choose how we want to see an electron.

However, there is a difference between bullets fired from a gun and a stream of electrons passing through one slit. A bullet fired from a gun follows Newton's equations; using them, it is possible to predict exactly where it will hit the wall and when. This is not the case with an electron; quantum theory stipulates that it is *impossible* to calculate the trajectory of an electron and to predict exactly where and when it will hit the wall. The theory represents an electron or any other particle as a superposition of an infinite number of waves, each with its specific wave length, *the wave packet.* Each of the waves represents a possible state and its probability of being actualized. The German physicist Max Born (1882–1970) showed how such probabilities could be calculated: they are proportional to the amplitude of the waves. Therefore there is no such thing as the trajectory of an electron. It is possible to speak about electronic density but not about the coordinate of an electron. For instance in our former experiment, the equations describing the electron allow us only to calculate the chance the electron has of hitting the wall at a specific time within a specific area, given initial conditions. If the electronic density is low this probability is low, and if the electronic density is high the probability is high.

What is so revolutionary in this theory is that—unless it is being affected by the eyes of an observer—an electron is a superposition of all the many possible states, the many possible options, the total probability of which is one. In the experiment above, before we try to detect it, an electron is everywhere; it is spread all over the space, it has no boundary. But as soon as we look at it the electron hits the wall at a specific spot. There is a momentary certainty about its location. All the other options vanish, meaning that the probability of finding the electron elsewhere is zero. This is called the *collapse* or the *reduction of the wave function*. The electron momentarily transforms into a located entity, into a particle.

The best image that comes to my mind is very far from the truth but can give a hint of what is meant by the *collapse of the wave function*. Imagine a huge invisible balloon the size of the whole universe. This is, in fact, the volume that one single free electron can take, in theory. This huge balloon is filled with the "presence" of the electron. The electron is not located at any certain point. However if a tiny hole is made in the invisible membrane of the balloon, the electron will pass through the hole, as if it were "collapsing." Observing an electron could be imagined, as a first approximation, as making a hole in the balloon.

We could resume our former experiment, but the mystery only grows: If electrons pass, not as a stream but *one by one,* through the thin wall with the two slits, again an interference pattern is observed! It is as if one electron could interfere with itself! Moreover if electrons are passing one by one through the wall in such a way that *we can know* which slit a single electron went through, the interference pattern disappears: the electron acts like a tiny bead!

The great physicist Richard Feynman (1918–1988; Nobel laureate in physics in 1965) pointed out that the entire mystery of quantum mechanics is in this double-slit experiment! Other more sophisticated experiments reveal even odder behavior of particles, such as entanglement, in which it appears that particles generated during the same event are connected even when separated by hundreds of thousands of miles. For those interested, a few physicists, like Fred Alan Wolf, have described these strange quantum observations at length in many books accessible to the general public.[3]

This odd behavior of particles should reminds us of what David Gross (Nobel laureate in physics in 2004) said one day at a conference: "Our knowledge enables us to evaluate the extent of our ignorance." In this case, ignorance is inherent in quantum theory itself. It is expressed in the uncertainty principle, first formulated by the German physicist Werner Heisenberg (1901–1976). This amazing principle is one of the pillars of quantum theory. In brief, it stipulates that it is impossible to know simultaneously both the precise location and the exact velocity of an electron, that only one of these parameters can be defined with certainty. The uncertainty principle also stipulates that we can never know the exact energy of an electron at a specific time. Let's put it this way: if it were possible to know precisely both the location and the speed of electrons or their energy at a specific time, electrons would not be part of this world; in other words, without this uncertainty in our knowledge, there would be no atoms! We would be erased from this world like spies who know too much!

In the quantum world, the observer plays a crucial role and influences the outcome of an experiment. Before an experiment, an electron is spread in the whole space, carrying endless possible states, the probability of which can be calculated. However, during an experiment, only one of these probabilities is realized. In other words, observation momentarily freezes the electron in only one of its many different possibilities. Undeniably, the electron is not the same as it was before the observation; it is in one state, only momentarily and only in the eyes of the observer. It is as if observation extracts the electron from the quantum world in which it possesses all its possible states, and installs it momentarily in our world, fixing it in only one of these states.

Here is a familiar image that might make this strange quantum phenomenon more comprehensible. Imagine yourself to be slightly like an electron, that is, with many possible options concerning your location: for instance, this evening you can either be at a concert or in the house of friends or at home in front of the TV. In your imagination you can explore each of these possibilities, allotting to each a probability of being realized. You can even imagine that this evening you will be in Tahiti on a beach, but if this option is not realizable, its probability is zero. By the end of the evening, one of the options will have been implemented and all the others will have vanished. The "observation" in the electron's case would correspond in your case to moving from the imaginary to the real. As with the electron, only one of the states that you had in your mind became real; all the other possibilities vanished.

Obviously, after one of the possibilities has been implemented, you are no longer considering the other options, and you are no longer exactly the same person as before. We could imagine that since we are made of an enormous number of these particles, all of which "carry" with them many possible states (theoretically, an infinite number of states), we also "carry" with us many possibilities, many options. In this quantum world of ours, who is the observer? Who chooses?

These extraordinary quantum phenomena and their obscure mechanisms have aroused much debate. The different interpretations offered split the scientific community. Even de Broglie was not satisfied with the interpretation of these "waves." In 1966 he wrote with irony: "By which strange coincidence, a representation of probability could propagate in space in the course of time, like a real physical wave likely to be reflected, to be refracted, to be diffracted?" Einstein thought that quantum theory was incomplete and that some elements were missing, which, if discovered, would modify this extravagant image of matter. Einstein criticized the dominant role of probability in quantum theory, saying, "God does not play dice." He viewed quantum mechanics as a useful stage between traditional physics and a future physics. He wrote to his friend Niels Bohr (1885–1962; a physicist from Denmark and a major figure in the quantum revolution): "I am not satisfied with the idea that we possess machinery that enables us to prophesy but to which we are not able to give a clear sense."

Einstein's remark is reminiscent of the comment of the Polish astronomer Nicolaus Copernicus (1473–1543) on the geocentric system of the Greek astronomer Ptolemy (160–125 BCE), who developed a system that made it possible to calculate the eclipses of the moon, the sun, and the conjunctions. This system was used successfully by astronomers for more than fourteen centuries, until Copernicus overturned it. Ptolemy's system was founded on monumental errors: that the earth was immutable, stood at the center of the universe, and that all the stars revolved around it. Copernicus wrote, "Such a system looks to me neither complete nor agreeable to the mind."

Today the majority of scientists think that the quantum theory is complete. However, they remain uncertain about many troublesome questions. For example: if, as some think, our own consciousness plays a paramount role at the time of the observation of quantum phenomena, what can be said of measurements made with eyes closed or at a distance? Original answers have been proposed, but all attempts to explain the quantum world encounter serious problems and put into question fundamental concepts like space and time; some even claim the existence of parallel worlds. Trying to explain quantum phenomenon seems nearly as hazardous as opening Pandora's box.

Many scientists accept the fact that a theory cannot provide an exact description of the world; it is enough for it to provide a model that can be used to make accurate predictions (like Ptolemy's) and that opens up new fields of research. Quantum

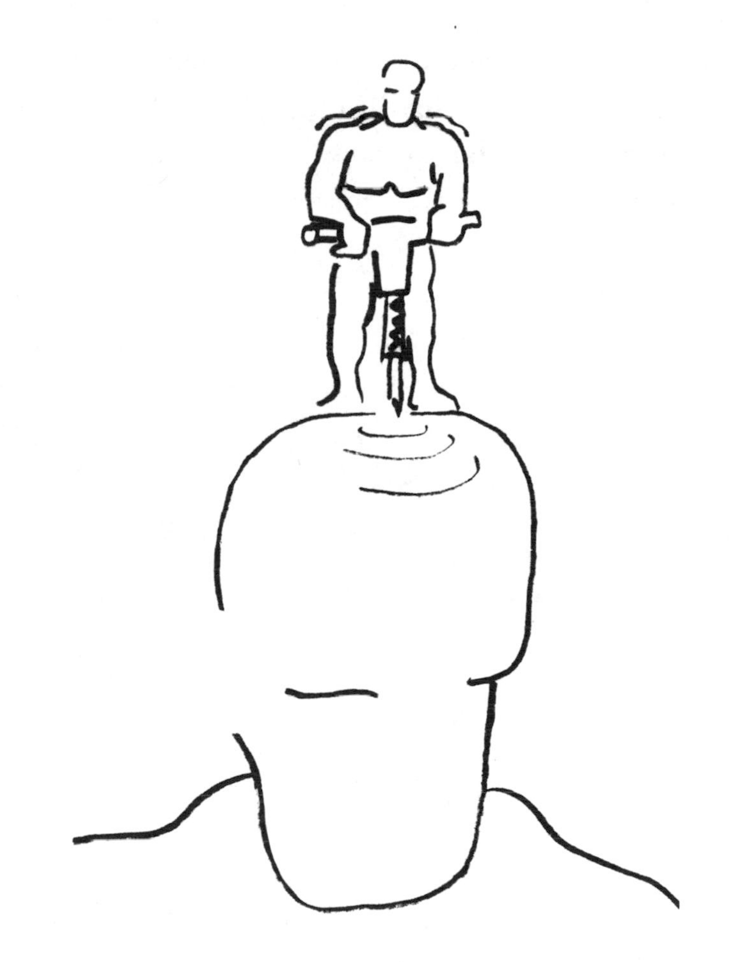

Figure 12.3. Matter is indivisible from mind

theory certainly fulfills these requirements. Does one need to know how a satellite really works to watch *The Simpsons* on TV? Or to predict that the neighbors will see it too if they turn on their TV at the same time and the same channel? According to de Broglie's equation, the wavelength associated with macroscopic objects is too short to be detected by any existing equipment. Therefore most scientists believe that whatever the physical reality of this wave/particle mix, it has little relevance for daily life, and they prefer not to deal with it.

Some scientists, however, found the idea of the wave/particle duality as uncomfortable as the idea of the body/mind duality is to specialists of our unconscious consciousness. One of them, the physicist David Bohm (1917–1992), proposed a striking image: he developed the theory that matter was an entity with two aspects, a physical aspect and a mental aspect; the mental aspect would correspond to what is most subtle and the physical aspect to what is the least subtle. According to his idea, matter is indivisible from mind, just as a particle is indivisible from its wave character. In short, he saw the particle and its wave character as a body and a thought. What would a particle think about?

PART THREE

When Reality Becomes a Choice

13

COMMUNICATION BETWEEN MOLECULES

As noted in the previous section, a living organism is an informed chemical system. As long as it lives, it contains an enormous amount of information generated and transferred throughout the whole system. Thanks to this information, an organism is able to feed itself and reproduce. Although we do not control or even understand most of it, this information is unambiguous to the system. Your molecules do not need a degree in biology or quantum mechanics to decipher it. On the contrary, the dream of many biologists, I reckon, would be to know as much as their own molecules do.

In the current model describing the principal mechanism of the exchange of information between biological molecules, it is believed that information is written in the structure of these molecules. As soon as a signal is given, the molecules start to react. They start to persistently "touch" one another, looking for a "mate." When a key molecule finds its lock molecule, a reaction is triggered, the product of which will also be a lock or a key, that is, a molecule that will trigger another selective reaction and produce other molecules. These will in turn also trigger reactions, until the last molecule of the chain is fabricated, that is, until information reaches its final destination. Although this model has been greatly improved in recent years, it still appears a bit clumsy and does not describe how a molecular structure could store information.

In the first part of this book, using diamond and graphite as examples, we described the determining impact of geometry on the properties of a molecule and its reactivity. Therefore the study of molecular structures is very important for all researchers, biologists or not, and is the subject of a vast area of research. No doubt many of you would be surprised to learn that ethanol, the alcohol found in wine and all other alcoholic beverages (see figure 13.1), has exactly the same chemical composition as dimethyl ether (see figure 13.2). In each case their molecules are composed of one oxygen atom, two carbon atoms, and six hydrogen atoms.

However, as you can see in these figures, the position of the atoms in each of the compounds is different; as you might guess, the properties of these two compounds are quite distinct. Dimethyl ether is a very volatile liquid that, unlike ethanol, cannot be swallowed or even inhaled without danger. In other words, more than a chemical formula is needed to identify a compound. Its structure must also be known.

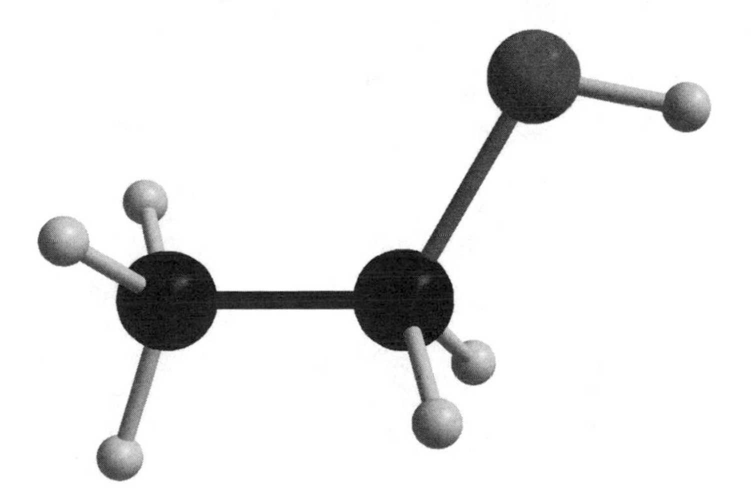

Figure 13.1. A computerized image of a molecule of ethyl alcohol: one oxygen atom (top right), two carbon atoms, and six hydrogen atoms

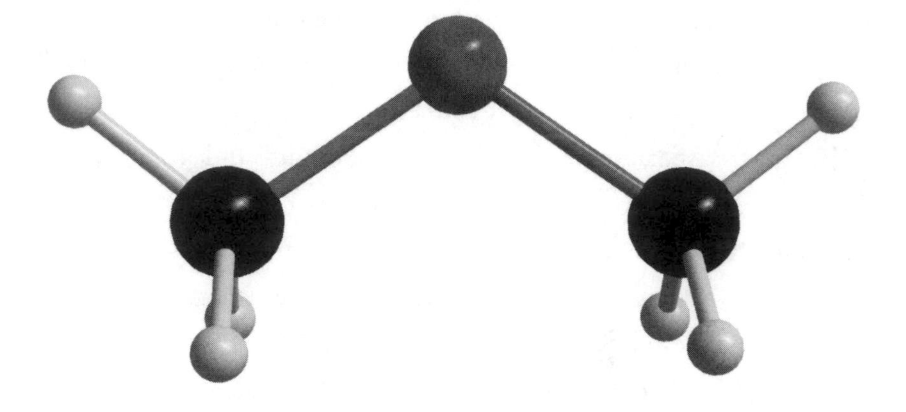

Figure 13.2. A computerized image of a molecule of dimethyl ether: one oxygen atom, center; two carbon atoms on the left and right; and six hydrogen atoms

But how do we know molecular structures? How do we know, for example, that graphite is composed of long layers of carbon laid one on top of the other, whereas diamond is composed of small, adjacent pyramids? How do we know what a water molecule looks like? Or a sugar molecule? Even if I tell you that a sugar molecule like glucose (a sugar that comes from grapes) is made up of twelve carbon atoms,

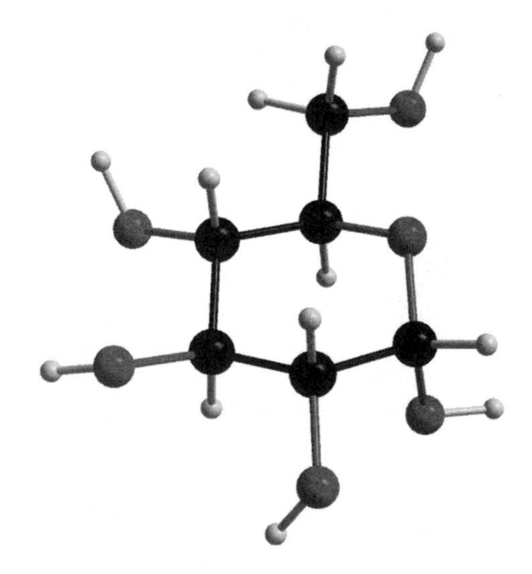

Figure 13.3. Computerized image of a molecule of glucose

eleven oxygen atoms, and twenty-two hydrogen atoms, they could be arranged in tens of ways. Only one would be glucose (see figure 13.3).

Another arrangement would correspond to galactose, a totally different compound that is much less sweet (see figure 13.4).

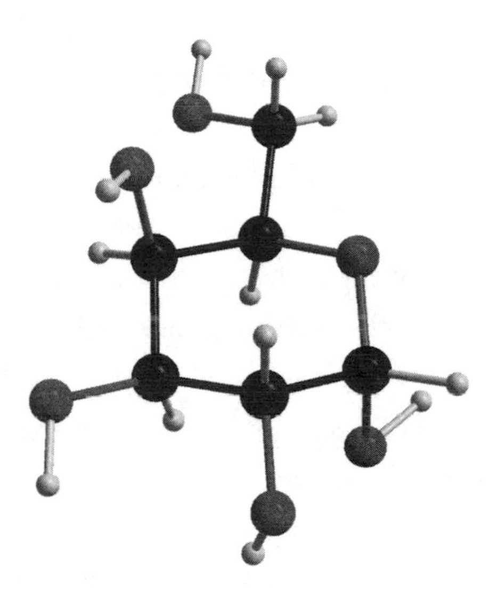

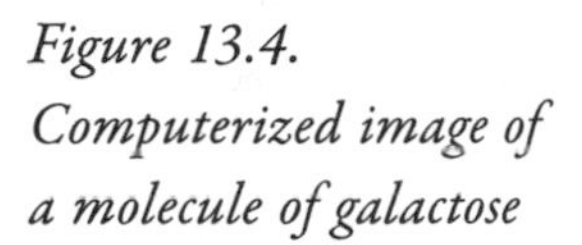

Figure 13.4. Computerized image of a molecule of galactose

Today scientists can know the molecular structure of a compound, if its molecules are sufficiently well packed and ordered like eggs in a box. Arrangements of this type exist in crystals. In 1912 the German physicist Max von Lave (1879–1960) discovered that when X-rays hit a crystal, they are diffracted, as if each of the irradiated atoms were acting like a tiny prism. The diffraction pattern they produce contains a lot of information about the relative position of the atoms in the molecule. Analysis of this information makes it possible to reconstruct the three-dimensional structure of a molecule, almost as if it had been photographed from all angles. In short, we can say that with proper lighting (X-rays) and under certain conditions (when the observed molecules are well-ordered) we can "see" the structure of a molecule. In this way, scientists have succeeded in deducing the "image" of all kinds of molecules, from small ones to very large ones.

In 1962, the molecular biologists Francis Crick (UK), James Watson (USA), and Maurice Wilkins (UK) were jointly awarded the Nobel Prize for physiology or medicine. Through their brilliant analysis of the diffraction pattern produced by X-rays hitting DNA, they succeeded in making a good picture of this huge molecule. In 2009 the Nobel Prize for chemistry was awarded to Ada. E. Yonath (Israel), Thomas A. Steitz (USA), and Venkatraman Ramakrishnan (USA) for having revealed the structure of ribosome, another crucial biological molecule. The pictures of the molecules "taken" by scientists all over the world are deposited in enormous data banks accessible to all, and several hundred new pictures are added every week.

However, for a few decades now, thanks to computerized simulations, it has been possible to guess the structure of certain molecules without using X-rays. It is enough to feed the computer with the chemical composition of the molecule (the names and number of its atoms) and a few assumptions. The computer calculates all the possible geometrical arrangements and deduces the most plausible one, like a detective constructing a suspect's identikit portrait from witness testimonies. The precision of these images, that is, the degree to which these identikit pictures resemble the "real pictures" of molecules deduced from the experiments with X-rays, is often quite impressive.

Chemists frequently use such computer programs to predict or confirm experimental observation or to see if a particular combination of atoms could exist. For example, could a molecule containing three calcium atoms (Ca), five chromium atoms (Cr), and four bromine atoms (Br) be stable? The results help the chemists to determine whether they should try to make such a molecule. I assume that, in this case, the computer would say no, but that's not important. What is important is that the equations used by the computer for these calculations are the fruit of quantum theory. The computer is able to predict a molecular structure through calculations based on the equations describing the wave packets of the molecule. If a molecular structure contains information, and this molecular structure can be "constructed" just by using the mathematical functions describing its wave packets, we must conclude that this information has its source in these wave packets. In other words, the information that keeps you alive and makes you a unique human being is stored in the wave packets of your atoms and your molecules.

We tend to believe that the information stored in the wave packet of a particle can be retrieved and understood only by specialists of molecular biology or quantum chemistry. But, in fact, as previously mentioned, every single molecule of a living organism understands it. Each single molecule of your billions of billions of billions of molecules "deciphers" this molecular information in its own way; the result is that you are a chemical system that knows how to keep its integrity, feed, protect, and reproduce itself. However, if information has its source in atoms and molecules, information exists in all molecular systems and not only in living systems. What, then, is the difference between living matter and inert matter? What is the difference between the molecular information stored in a bacterium (what we call biological information) and that stored in a pound of salt?

To suggest an answer, let's learn a *bit* about information: when we think about the quantity of information that we process each day, whether through television, radio, telephone, the Internet, newspapers, movies, and so on, it is hard to conceive that information can be weighed like diamonds. And yet, in the middle of the twentieth century, a young mathematical genius, the American Claude Shannon (1916–2001), revolutionized the world of telecommunication by developing a new theory, *information theory*. He defined a new unit, the *bit,* and paved the way to the digital revolution.

If weight is measured in pounds, distance in miles, and volume in pints, information is measured in bits. The theory states that the quantity of bits contained in information—we could say the "weight" of the information—is inversely proportional to the amount of uncertainty it contains. In other words, the more probable the information, the lighter it is in bits; the more improbable the information, the heavier it is in bits.

If, for example, an announcement is made through the media that tomorrow the sun will rise in the east, the quantity of information contained in this proclamation is zero, because it is quite certain that tomorrow, as on any other day, the sun will rise in the east. However, if an announcement were made that because of certain phenomena observed by satellites in space, and according to calculations made by NASA scientists, next Monday an enormous meteorite will fall on and crush the White House, the amount of information contained in this announcement would be enormous, and would certainly overwhelm all media outlets. It is enormous because it is totally unexpected, and it is totally unexpected because it is very improbable.

We could imagine information decreasing or increasing as it becomes more or less precise. Let's take, for example, information about the weather on a summer day: "It is raining today in London." The amount of information contained in this message will decrease if the statement becomes less precise and more probable: "It is raining today in England." And it will increase if it becomes more precise or less probable: "It is raining in London this morning," or "It is raining in Cairo." The more probable an event is, the less information it contains; inversely, the less probable an event is, the more information it contains.

Let's return now to chemical systems and to our question: What is the difference between information stored in living matter and inert matter? We might be tempted to reply now that since, according to information theory, an improbable phenomenon contains much information, the difference is that a living organism contains more information than inert matter, because, dwelling far from equilibrium, a living organism is an improbable phenomenon. However this is not the full answer. Since information is contained in the wave packets of all atoms, the more atoms there are in a chemical system the more information that system contains. In other words there is probably more information in a pound of salt than in a tiny bacterium. Therefore, even though a living organism contains much information, there is more to *biological* information than its quantity.

You are a molecular system in constant activity. *Some* of your molecules are heading toward life; the rest are heading toward death. Every day, several billions of your molecules are ejected from your organism and make their way to equilibrium through the gases, the liquids, and the solids you expel as you go about daily life. On the other hand, new molecules are constantly being created, thanks to energy and matter that you absorb from outside. We have associated these two processes with "the instinct for death" and "the instinct for life," respectively. Indisputably, the replication of your cells is no more essential to your existence than the expulsion of dead cells. We can say here, not without irony, that death is vital. It is this constant renewing of old molecules by new ones that keeps your system in constant activity, that is, far from equilibrium and alive.

Chemically speaking, you are not the same as you were five minutes ago. Therefore you are not obliged to panic and take medicines every time you feel bad; it might pass. Unlike table salt, you are a reacting chemical system, the molecules of which are temporary. This is another privilege of living matter: most of its molecules are in transit. They are constantly being renewed. This constant activity keeps your whole system in a metastable state. At any given moment, some of your molecules are moving toward equilibrium while others remain far from equilibrium; but there are enough of them remaining far from equilibrium to keep you alive. (If all your reactions tended toward equilibrium, all your cells would decompose and you would rot in few days.)

As we know only too well, this situation is not permanent. Sooner or later a living system is transformed into an inert chemical system. There comes a moment when your molecules, still full of information, change direction and begin to move toward equilibrium. On the way they lose the information they contain, which is necessary to maintain the integrity and the molecular activity of the chemical system that is you. Sometime after your death, practically nothing remains of the information that made you a whole and unique chemical system. Long fragments of your DNA, if they have been given the opportunity to do so, will remain for a time in other living chemical systems—those of your descendants. The remainder of your molecules, made up of your imperishable atoms, will be launched into the huge molecular network that is our universe. There, other destinies await them, through random encounters like those so poetically described by Primo Levi in the last chapter of *The Periodic Table:*

> I could recount an endless number of stories about carbon atoms that become colors or perfumes in flowers; of others which, from tiny algae to small crustaceans to fish gradually return as carbon dioxide to the waters of the sea, in a perpetual, frightening round-dance of life and death, in which every devourer is immediately devoured; of others which instead attain a decorous semi-eternity in the yellowed pages of some archival document, or the canvas of a famous painter; or those to which fell the privilege of forming part of a grain of pollen and left their fossil imprint in the rocks for our curiosity; of others still that descended to become

> part of the mysterious shape-messengers of the human seed, and participated in the subtle process of division, duplication, and fusion from which each of us is born. . . .
>
> One, the one that concerns us (the carbon atom), crosses the intestinal threshold and enters the blood stream: it migrates, knocks at the door of a nerve cell, enters and supplants the carbon which was part of it. This cell belongs to a brain, and it is my brain, the brain of the me who is writing; and the cell in question, and within it the atom in question, is in charge of my writing, in a gigantic minuscule game which nobody has yet described. It is that which at this instant, issuing out of a labyrinthine tangle of yeses and nos, make my hand run along a certain path on the paper, mark it with these volutes that are signs: a double snap, up and down, between two levels of energy, guides this hand of mine to impress on the paper this dot here, this one.[1]

I wrote earlier that your molecules will lose the information they contain, but this is not exact. Only biological information is lost, not chemical information. Molecules cannot lose information since the latter is engraved in their wave packets, that is, in their nature. So, again, what is the difference between biological information and chemical information? What is lost when a living organism dies?

My intuition is that information stored in the wave packets of molecules behave also like waves. And like any other waves, like the waves of the wave packets, these waves—information waves—interfere. They can be coherent or incoherent. In fact, I think that matter waves do not *store* information but rather *are* waves of information. One can imagine that biological information corresponds to a long-range coherence between these matter waves or information waves, and what is lost when an organism dies is coherence. In other words, I believe that we are alive thanks to quantum coherence at a macroscopic scale. Compared to inert matter, which exhibits quantum coherence only at the atomic scale, you and I and all that is alive exhibit quantum coherence at a macroscopic scale.

To conclude, I suggest that what makes the bacteria (but not the pound of salt) alive is not the quantity of information but its quality. In salt and in all inert matter, it seems that the wave packets are coherent only at the atomic scale, where they generate the chemical bond between atoms, but in a living organism, coherence might happen on a much larger scale. It seems that being far from equilibrium is a potential for life because there is a potential for an improbable phenomenon: coherence of the wave packets at macroscopic scale. This is quite improbable because it implies that many molecules behave as if they were "tuned." Indeed, if we think of the size and number of molecules in an organism, this sounds improbable, but not impossible. In fact long-range quantum coherence has already been observed by scientists in superfluids, which are fluids with zero viscosity. Though quantum theory deals with matter on the scale of atoms and atomic particles, at low temperatures (for example, at less than −455°F, liquid helium becomes a superfluid) superfluids exhibit phenomena that are manifestations of quantum behavior on a macroscopic scale. Superfluids all have the unique quality that their atoms are in the same quantum state. This means that if one moves, they all move in the same direction. I suggest that molecules far from equilibrium also have a chance to exhibit long-range coherence, to communicate, as Prigogine wrote.

In summary, matter is made of molecules. We tend to see molecules as solid particles, but—in its intrinsic nature—any particle of matter is a wave packet, which contains information. With each reaction, inert systems spontaneously move toward equilibrium, the most probable state, and information gets lost, whereas a living system can move far from equilibrium—an improbable state—and aquire information. Actually information doesn't get lost, just as matter cannot be lost, but it may become incoherent, "unreadable," like a word, the letters of which that have been mixed up.

In fact, any chemical system can be conceived as carrying information stored in its wave packets. During a chemical interaction, these wave packets interfere, causing information to become more or less organized or coherent, depending on whether the chemical system is tending away from equilibrium or in the opposite direction. We might assume that the attainment of a certain amount of coherent information could generate "communication" between molecules. This mode of communication is not based on the structure of a molecule and would be appropriate to huge biological molecules as well as to very small molecules like those involved in a Belousov reaction.

Today we believe that living beings are the product of a long process of evolution of inert matter. To describe this long process, one can imagine that in a molecular network condemned to a slow journey to equilibrium and molecular chaos, statistical-outlier molecules escaped this fate. Far from equilibrium, these molecules were able to spontaneously organize into the first dissipative structures. Prigogine asserts that this organization would be conceivable only if molecules communicated with each other. This communication can be imagined as long-range coherence between the wave packets of these molecules. From the beginning of time until the appearance of life, that is, for a billion years, this communication would have developed and dissipative structures would have grown and evolved, storing more and more coherent information, until they became so rich in information that they were able to reproduce.

Your own molecular system developed from the fertilization of an ovum by a spermatozoid, in other words, from the meeting of two molecules of DNA (or two wave packets), that of your father and that of your mother. Their affinity started a chemical reaction that, thanks to the information these molecules already contained, involved more and more molecules. This process, during which information increased, led to the creation of your first cell. That meeting could have had no consequences, or it could have involved another of the 200 million spermatozoids provided by your father. In any case—by an extraordinary chance—you won the race against equilibrium and *your* first cell emerged. This cell produced many other cells that organized themselves into an increasingly sophisticated and voluminous molecular system. When it became too large it was expelled from the system where it had developed so far, your mother's womb, and you were born.

You were stuffed with information, sufficient to enable you to continue to develop on your own, or nearly on your own. In this immense network of molecules where you found yourself at your birth, you were a prodigious phenomenon. While nearly everything that moved tended implacably toward increased molecular chaos, you appeared and evolved in the opposite direction, as if turning your back on this growing chaos. You did it because—as the opposite of an inert chemical system that tends toward increased molecular chaos and death—your molecules tend toward life, where information increases.

You can imagine your lifespan as the more or less time-consuming journey of *all* your molecules toward equilibrium. This course would coincide with the gradual decoherence of the wave packets leading to the gradual loss of the molecules' ability to communicate. As if the dialogue between your cells slowly diminishes as you grow older, the messages that they transmit become less and less coherent, less and less clear and comprehensible. The end of your life could be likened to stopping the construction of the tower of Babel because the workmen were no longer able to communicate with each other.

Could the evolution of which Oparin wrote, of inert to living matter, correspond to increasing communication between molecules? Through long-range coherence of the wave packets? Has this communication evolved since? Can evolution of communication be identified in the evolutionary chain from protozoan to human? Can the appearance of language be seen as a phase in this evolution? Has this evolution ended? Is the invention of printing, radio, telephone, and so on, and, more recently, the Internet, which transforms our life every day, also a phase of this evolution?

Erich Fromm (1900–1980), the German humanistic psychoanalyst, wrote: "The major need of man is to overcome his separation, to flee the prison of his loneliness. . . . The man of any age and any culture is confronted with the solution of a single problem: how to overcome separation."[2] Is this need for communication engraved in our molecules? Like hunger and thirst?

14

THE HOLOGRAPHIC UNIVERSE

Now we are going to venture further beyond the borders of science and delve into areas that many scientists refrain from exploring. Do not worry too much. Don't forget that modern science is limited to the study of phenomena that scientists know how to measure, and that we have no reason to believe that certain phenomena scientists are unable to measure today will not be measurable tomorrow.

I assume that many of us are ready to agree that a young athlete at the peak of his career possesses more life force than an old, disabled, and sick man. Yet life force, as already mentioned, is not quantifiable, and therefore this term does not appear in the scientific lexicon. This doesn't mean however, that a life force will never find a place in an equation. We can take another example: our thoughts. We all experience different kinds of thoughts: black thoughts, pink thoughts, clear thoughts, invading thoughts, passing thoughts, and so on. But still, our thoughts are not quantifiable. We cannot assign them a weight, a volume, a color, and so on, yet we don't conclude that we do not think.

Let's take a last example, which we have already talked about at length. According to quantum mechanics, any particle of matter behaves like a solid body in some cases and like waves in other cases.

This depends on the setup of the experiment or, in other words, on the observer. However, this does not mean that a particle of matter is a body *or* a wave, or a body *and* a wave, but that sometimes it can hit a wall as a solid body does and sometimes it produces interference, as waves do. However, today we can't observe the interference patterns produced by macroscopic objects.

The first report of an interference pattern produced by matter was made only a few decades ago by Davisson and Germer. Since then, scientists have observed interference patterns produced by whole atoms, and recently it has been possible to observe interference patterns produced by fullerene, a compound made of sixty carbon atoms.[1] However, detection of interference patterns coming from more sophisticated or larger molecules is impossible with current technology. This does not mean that in the twenty-first or the twenty-second century we won't succeed in detecting interference patterns produced by our own cells.

It is pertinent to mention the work of Jacques Benveniste (1935–2004). In the seventies Benveniste was one of the most published French scientists in immunology, until he began to research homeopathy. He was not interested in the therapeutic effect of homeopathy but in homeopathy's general claim, which seems to contradict common scientific sense.

A homeopathic remedy is prepared from a water (or alcohol) solution in which an active compound, such as salt, is dissolved into the liquid. It is then further diluted through a stepwise process in which the solution is vigorously shaken between each subsequent dilution. *Highly diluted* means that the concentration of the active compounds is below one milligram per liter—a concentration in which the number of molecules of any compound is very low. At some of the very highest dilutions currently used in homeopathy, there are no more foreign molecules at all in the water (or alcohol).

It is as if you took a grain of salt and put it in an Olympic swimming pool filled with clean water and then drank a glass of water from it. Would you taste the salt or feel the action of the salt in your body? Homeopathy claims that the higher the dilution, the stronger the effect of the active compound on our body! In other words homeopathy claims that even though there are no more molecules of the active compound in the water, the water keeps in its memory the souvenir of the molecules that were formerly dissolved there, the trace of their passage, their "footprints."

Benveniste's results were totally unexpected, for him too. They showed that there was truth in this very strange homeopathic claim. Benveniste published these results and launched the concept of the Memory of Water. However he could not reproduce his results at will, and his colleagues banished him. He quickly found himself isolated from the scientific community. Nevertheless he continued recording information from highly diluted aqueous solution like one records a piece of music. He died at the early age of sixty-nine. However though Jacques Benveniste has been buried and much too soon, his results and his ideas were not. In 2011, less than a decade after his death, Luc Montagnier, a French virologist and 2008 Nobel laureate in physiology and medicine, praised Benveniste.[2] Montagnier explained that today he believes Benveniste. His own results do suggest that traces of certain biological molecules could be detected and "recorded," like Benveniste's did.

We can therefore cross these artificial and probably temporary limits without too much apprehension, to enter the zones where scientific constraints are no longer obstacles to our imagination. Einstein himself said that imagination is more important than knowledge. According to legend, he conceived the theory of relativity by imagining that he himself was a photon of light; this revolutionized physics long before his theory could be verified, which did not occur for another fifteen years. Einstein discovered the theory of relativity by imagining himself simultaneously at the researcher's position, at the subject's position, and at the place of the measuring device.

As we have seen, information is written in every particle of matter, in every atom, every electron. We therefore have no reason to believe that information transfer involves only biological activity. In fact we are part of an immense network of information coming from a furious chemical activity, that of living matter in constant motion and that of inert matter, also constantly being transformed, whether by the smallest atmospheric change or by our own manipulations. In each chemical reaction energy flows and so does information. How? We believe information flows through interference, since these particles are strange entities, which, when we do not observe them, behave like waves. So when no one observes us, not even ourselves, our billions of billions of billions of particles "regain" their wave aspect.

I therefore propose you close your eyes and imagine these wave packets of yours. It will not be foolish to imagine that you are alive thanks to the long-range coherence of the wave packets of each of our molecules. As long as you are alive, coherent information traverses your organism constantly, from top to bottom; it keeps your integrity and your specificity. It animates your body and makes your molecules work in unison. If your molecules' wave packets were not coherent and were destroying each other, your system would probably decompose and move toward equilibrium, that is to say, toward chaos and death.

In fact our whole universe seems to be a network of waves, the waves of all the wave packets of the molecular soup. Your wave packets represent a tiny but unique part of the infinite network of information that is our universe.

Thanks to your five senses—sight, hearing, smell, taste, and touch—you can perceive the rest of the universe. Like five doors, these senses represent your sole access to the outside world and your personal connection to it. When one of your senses is excited it triggers multiple chemical reactions, such as when we look at something—let's say a big red square canvas. In the dark you cannot see the canvas; so the experiment begins with the reaction between the canvas and photons of light. As regular light is composed of all the colors, we can imagine this light as a mixture of colored photons. When these photons, which we can imagine as tiny colored drops, hit the canvas, the molecules of the red paint act like sponges: absorb most of them but reject all the red ones. When these tiny red drops (the red photons) reach your eyes, they trigger a chain of chemical reactions at the end of which certain molecules are formed, carrying new information: "the canvas is red." It is as if in your organism there was a radio giving a news flash that we can "hear" from inside. Comparable chemical processes accompany all your sensations of seeing, hearing, smelling, tasting, and touching. They all trigger chemical chain reactions at the end of which you and your molecules get information, most of which you aren't aware of. In his article Sason Shaik reminds us that we are but a chemical factory.[3] For instance, he writes that when a man and a woman are attracted to each other, "the whistle blows" at the phenylethylamine department of their factory. Research suggests that phenylethylamine is a neurotransmitter and a neuromodulator of libido, believed to be associated in both sexes with the event of falling in love. If it happens that our protagonists become more intimate, in his factory the whistle will also blow quite loudly at the nitric oxide department. Nitric oxide (natural Viagra) will reach his penis, causing a cascade of chemical events by the end of which the muscle cells are depleted from calcium, causing the muscle to relax and thereby enabling blood flow into the penis and its eventual erection. In her factory, the production of oxytocin will suddenly rise and eventually cause her orgasm.

We usually grant our brain the role of a control room. We believe that our brain, this fantastic and complex network of neurons, ensures the efficient flow of information inside our organism. DNA sequences of one species are so similar to those of another species that mistakes might easily occur. Without some control, nothing would ensure that, for example, a woman who mostly eats dairy products would not start to low like a cow or give birth to a calf, or that we ourselves would not wake one day bearing fangs and a tail! It seems plausible to imagine that, of all our organs, our brain would be the one to do the job. It would gather information, then sort and distribute it, just as mail is sorted and distributed to the right address. There is nothing unusual about granting the brain this role. However, it seems that our brain is more than a simple postman or a traffic policeman.

Let's remember that the brain has been clever enough to conceive of both the radio and the fax; therefore we can expect it to be at least as sophisticated as these machines. A radio, for example, receives waves (electromagnetic waves) full of information, sometimes sent from abroad, and transforms them into sound. A fax machine also receives electromagnetic waves containing information sent from the other end of the phone line and transforms them into a text or an image, like a translator who translates Chinese into English.

As our whole universe is a network of information waves, we might imagine that our brain not only directs information but also "translates" at least a part of it for us. It would seem plausible to think that our brain translates the information coming from our visual system into shapes and colors, that coming from our auditory system into sounds, that coming from our olfactory system into smells, that coming from our gustatory system into tastes. Finally, our brain translates the information that strikes our skin into textures. It is not foolish to imagine that our brain acts as a very sophisticated device, a device that "reduces" all these wave packets of matter, all these waves of information that surround us, into definite shapes, colors, sounds, smells, tastes, and textures, just as the observer in the double-slit experiment reduces the wave aspect of an electron to a tiny solid body. Our brain would be the ultimate observer of the universe.

It is quite mind-blowing to realize that, from a scientific point of view, nothing prevents us from believing this. Nothing prevents us from believing that we and everything around us are only waves, waves of probabilities, and that when interacting with this very sophisticated device called our *brain* these waves are reduced to certainty. A certainty WE choose. Then specific information about colors, shapes, sounds, tastes, smells, or textures is released. With this released information, our three-dimensional reality is created. However, this reality has been created from waves, like a hologram. Our reality is, in fact, a hologram and our whole world a holographic universe. The idea of a holographic universe is widely spread among scientists who, like David Bohm, believe that the universe we see and experience is but a holographic image created by the mind. From the thermodynamics of black holes, the information relative to our whole universe, let's say its holographic film, could be "located" at the boundary of our universe. According to science we all are victims of a fantastic sensory illusion. How far does this illusion go?

15

MOLECULAR MOODS

Close your eyes and imagine your body, immersed in a huge network of traveling information. Imagine your five senses—sight, hearing, smell, taste, and touch—interfering with the information crossing this network. And imagine your brain as a device that sorts them and translates some of them in a language you know: colors, sounds, smells, tastes, and textures. So far this sounds quite easy to imagine. But does this translation signed by your brain end there? Do you receive information other than colors, shapes, sounds, tastes, smells, or textures?

Let us resume our former experiment. Close your eyes and imagine again that you are looking at a square red canvas. Thanks to the reactions triggered by the red photons in your eyes, your brain receives information from the wave packets of the red canvas, like a fax machine receives telephonic information. Your brain translates this information into color and shape: the canvas is red and square. Now pay attention. While looking at this canvas, do you receive information other than its color and shape? Wash your eyes well with the red drops coming to them, then take the red canvas away and replace it with a black one. What happens? How does it feel? Are your emotions the same in front of the red canvas and the black one?

Are your emotions the same in front of other canvases, like the *Water Lilies* by Claude Monet or *The Scream* by Edvard Munch or *The Birth of Venus* by Botticelli? If our emotions were the same in front of all of these paintings, then why did these masters bother so much to paint them? And what, after all, is the difference between a portrait painted by Picasso and the drawing of a child? If, however, your emotions are altered by different paintings, where does this change come from? What has changed but a very thin layer of molecules that covers the canvases? Could your emotions be related to this thin layer of molecules? Could your emotions be caused by the reaction of your own molecules?

We usually believe that emotions belong to the domain of the mind, because emotions bring back memories and lead to thoughts and dreams; they trigger imagination and can even invade our whole mind. But is it true? Is it true that emotions belong *only* to the domain of the mind? Could your emotions have their source in the activity of your molecules? Could your emotions also be information coming from matter? That is, information received at the end of a long chain of reactions, triggered by your five senses and interpreted by your mind? Are your emotions connected to matter too? Would your emotions be a link between matter and mind?

Mind is a word that has a very broad meaning. It is, therefore, not easy to talk about mind in a context that is slightly more scientific than usual. Nevertheless, it seems that mind has many aspects. One aspect of our mind might have a behavior similar to that of our molecules: it is in constant agitation and can, like our molecules, be transformed in the wink of an eye (especially that of a beautiful woman or a handsome man). I mean that aspect of our mind that we commonly refer to as our state of mind or simply our mood. Undeniably this aspect of our mind, our mood, is affected by our emotions. A sound, a word, a smile, a touch, or a look might change our mood in an instant.

If our mood is dictated by our emotions, and if these are affected by the reaction of our molecules, this means that our mood is dictated by the reactions of our molecules. Is it so? Let's investigate further. Is your mood affected by the scenery and the colors around you? Do you have the same emotional response whether you are closed up in your house on a gray and stormy day or lying on a beach under a brilliant sun? Do you respond the same way to a devastated landscape as you do to a plantation of flowering peach trees? I don't believe so. Nor do I believe that your molecules react the same each time.

When this molecular system that is you has been drastically modified—after an amputation, for example, or plastic surgery, or even simply by having your hair dyed—would your emotions the same? And if not do you think these changes have been triggered by something other than the reactions of your molecules?

Can exercise change our state of mind? Some say that the principal aim of disciplines such as yoga, tai chi, or aikido is to change our mind-set. And what is physical activity if not a boost to some of our molecules? What about music? Don't we say that it can change our mood? Don't some pieces of music sometimes soothe us when our mind is tormented? And how do we hear a noise if not via the chemical reactions triggered in our brain by the vibrations of the tympanum? Some believe that even what we eat has an impact on our mood. Do you think that some dishes could excite you? And some teas calm you? And what is food if not chemical products that maintain the activity of our cells?

Here is another striking example: antidepressants. No one can deny that antidepressants act on our mood. However, like any other drug, their role is to offer our molecules other partners and make them take other trajectories. How would a change in the direction of your molecules affect your mood? I am not even referring here to recreational drugs, those that are proscribed and that specifically affect your state of mind, such as cannabis, LSD, cocaine, ecstasy, certain mushrooms, and so on.

I believe our emotions are directly related to our molecules. Our mood, that is, our state of mind, could be "a facet of mind that touches matter"; our emotions can be imagined like a bridge or an interface between matter and mind. It is noteworthy that Sir Roger Penrose, one of the greatest mathematicians of our epoch, connects cerebral activity to molecular activity. He thinks that cerebral activity may be related to what he called a "coherent quantum activity on a large scale."[1] Could this coherent quantum activity on a large scale be responsible for our emotions?

We can imagine our brain as part of a huge network of interfering waves carrying information. Like a computer circuit that dictates the flow of electromagnetic waves—the current flow—to create a virtual reality, our brain would interfere with the matter waves carrying information, dictate its flow, and also interpret some of it for us, creating the reality we are in. This interpretation would not only include colors, sounds, shapes, smells, and textures, but also emotions. I use the word *interpretation* because each of our brains is different. Each of us not only possesses a unique set of DNA necklaces in each of our cells, including our brain cells, but also our network of neurons is different. In each of us it evolves differently from our birth to our death. It is affected by our personal experiences and traumas. We could imagine that, having all gone to the same school, the brains of most of us would know how to read in the molecular language that "the canvas is red." But still each of us would make a unique translation, connected to our own brain's unique style, which is connected to our genes and, according to many schools, the pattern left by strong emotions from the past.

The whole universe could be a field of information that living matter would rather constantly gather and organize and inert matter would rather constantly mix and destroy. Where does the information come from? If I have succeeded, as I hope, in avoiding the slippery terrain of metaphysics so far, I am not entering it now. I will however quote again Albert Einstein: "Religion without science is blind and science without religion is lame."

16

CHOICE

What Music Do You Play?

Let's go further in these zones where imagination prevails and let's accept that the activity of your molecules is responsible for your emotions. That is, your emotions are the product of your reacting molecules, and the reactions of our molecules dictate our emotions. However, we know now that the reactions of your molecules (like those of all living matter) can move in one of two directions: toward equilibrium, meaning toward death, the most probable state; or toward the improbable, far from equilibrium, meaning toward the uncertainty of life. We could see ourselves as the constantly modified outcome of reacting molecules or wave packets, heading toward one of two possible directions: toward equilibrium or against.

Doesn't this remind you of a computer? A computer is also a machine through which waves are traveling—electromagnetic waves. These are moving through the computer, following a specific circuit. The circuit is so designed that at each node there are only two options: plus and minus. This apparent simplicity doesn't prevent this machine from generating a virtual reality that amazes us. I am making the comparison to point out that the apparent simplicity of the motion of our wave packets does not, in this model, limit the variety and the complexity of the information our system can contain.

Nevertheless, if your emotions are also a product of the reactions of your molecules, this has a fantastic consequence. This means that your emotions might also head toward death or toward life, according to whether the molecules involved in these reactions tend toward or away from equilibrium. In other words, if your emotions are also a product of the reactions of your molecules, your emotions affect your health.

For the time being, mainstream scientists deny it. Yet more and more voices are being heard saying it. Ancient traditions, mostly coming from the East, have succeeded in changing our Western way of looking at our body and mind. These traditions all teach that there is an intimate relationship between our mind and our body. They teach that each affects the other: a healthy mind in a healthy body. Some physicians—like Deepak Chopra, a pioneer in the field of mind-body medicine—have revolutionized the conventional medicine they learned and succeeded in shaking public opinion.

Could the close relationship between mind and body be mediated by our emotions? If indeed our emotions come from the reactions of our molecules, as we have suggested earlier, the question is: is this a two-way relationship? Could our emotions affect and direct the movement of our molecules? Could some emotions bring us nearer to death and other to life? Could some emotions act on us like a medicine? Others like a poison? Practitioners of feng shui, an ancient Chinese teaching, believe that the decor and colors that surround us are important to our physical and moral health. Practitioners of acupuncture, homeopathy, naturopathy, and so on all believe that emotions, like anger, for instance, might be so hurtful that they could work like a virus inside of us, with the ability not only to infect us but also those around us. Other emotions, like fear or frustration, might kill us slowly, like the bite of a poisonous snake.

Many schools are open today in which the main teaching is based on ancient traditions like Judaism, Buddhism, and Sufism, in which a main tenant is that our emotional state directly affects our health.

For example, they tell us that by controlling anger rather than letting it explode and cause unpredictable damage, or by seeking the pleasure in life rather than dwelling on its misery, we can slow down or even totally suppress destructive inner mechanisms. In short, they teach that hate and resentment, for instance, can be as hurtful to our health as venomous mushrooms or rotten meat, and that emotions like love and joy can cure us from all pain.

They also teach that emotions can be controlled and the power to control them has been given to all of us. These schools teach that we can choose our emotions; we have the choice. We have the choice to be angry at a bad driver or let it go; we have the choice to stick to frustrating emotions like jealousy, guilt, and envy . . . or not. Good schools do not tell you what to choose. They just teach you to recognize this choice. In fact this might be the only choice we have.

As surprising as it may first seem, there are very few things we can control or choose throughout our lives. Did you choose your date of birth? Your birthplace? Your skin color? Your parents? Your name? Your children? Your sex? There are many other things you haven't chosen that determine your life or can suddenly knock you off your feet. Are you protected from natural disasters? Epidemics? Accidents? Wars? Economic crises? Bad news? Or simply harmful encounters? Moreover, as part of a huge molecular network in constant motion, aren't we hit by billions and billions and billions of molecules of which we are unaware, which could turn us into a cripple or a vegetable in an instant?

In fact, most of the time we resemble a sailor condemned to spend his life at sea, at the mercy of the winds and the rain, of sharp rocks and encounters with other sailors and creatures of all kinds. Can he choose his course? To what extent did you choose yours? Without dwelling on this question, about which much ink has already been spilled, it seems that one of the few choices we can make, one of the few choices that is not a compromise, is the choice of our attitude to the events that shape every moment of our lives. Choosing an attitude is in a way choosing your emotions.

In summary, if emotions are not only a product of the activity of our molecules but also can change their course, our emotions and our molecules are, in fact, like the two sides of the same coin; or to use the image of the famous poet Khalil Gibran in *The Prophet,* like a rudder and a sail. As the rudder is inseparable from the sail, your emotions are inseparable from the movement of your molecules, and as the direction of your boat is sometimes dictated by the rudder and sometimes by the wind, the direction of your system would sometimes be dictated by your molecules and sometimes by your emotions.

Therefore I believe the duality we feel between mind and matter is an illusion. As dramatic as it sounds, the duality we feel might come from the motion of our billions and billions of molecules, going either toward chaos and death or toward life that breeds life, like an enormous swarm of flies caught between two jars of honey, the honey of suffering and death or the honey of life.*

I believe that through our emotions we affect the movement of our own molecules and their reactions. Moreover, as every chemical reaction involves an energy transfer, I believe that through our emotions we unknowingly direct a part of the energy flow in our body.

*The "honey of suffering" is an expression used by one of my teachers, Colette Aboulker-Muscat (1909–2003), to emphasize how attracted we might be to pain and suffering.

Ancient traditions claim it. I believe the only way we can direct our molecules, though quite limited, is through our emotions; we can choose to surrender to the certainty of death or to bet on the uncertainty life.

You are different from all that surrounds you. On this planet, there has never been a human being like you and there will never be another. You are unique, but you're not cut off from the world, even if sometimes you feel lonely. You are in constant commerce with the molecules that surround you. You are constantly being renewed and altered by the molecules that penetrate your organism, and certainly your presence constantly modifies this portion of your molecular network. Everything is connected to everything. In other words your presence doesn't go unnoticed in the eyes of the universe, even if sometimes you think otherwise.

If your emotions are closely related to the movement of your molecules, choosing your emotions might not only change the path of your molecules but the path of neighboring molecules. How far could this influence spread? In an article that made the cover of the *New Scientist* in 2009, Michael Bond explains how the friends of your friends can affect your mood.[1] In fact, the impact of our moods on our surroundings could have unsuspected dimensions.

If the information waves of molecules were making noise, like sound waves, we could imagine our universe like an orchestra hall during a performance of a symphony (some would say cacophony), the sounds of which would come from all the particles of our universe. You and I would be part of these constantly modified interference patterns. In other words, you and I would be part of the "symphony." If we are all unique instruments in a universal symphony, what kind of a piece do you play? How far do you resound?

17

MIND AND MATTER

Alchemists sought in matter much more than gold: they sought wisdom and the key to eternal life. Could those indeed be hidden there?

Today many feel that the state of our matter (that is, our health) and the state of our mind (that is, our mood) are closely connected. However, this idea, which intuition might sometimes be inclined to support, is generally ignored by scientists when not condemned outright. As we said, science deals only with phenomena that can be measured, and intuition is never part of an equation. Moreover many scientific mistakes, like the notion of the caloric, have been made in the name of intuition. Therefore many scientists prefer to ignore what cannot be measured; many others even deny the existence of what cannot be measured.

For most scientists today there is no connection between mind and matter. Mind belongs to the metaphysical world and matter to the concrete world. However, the discovery of quantum phenomena made all scientists agree that matter is, at the very least, an enigmatic phenomenon, and that the world as we see it has, in fact, no existence in itself. Indeed, the most enigmatic aspect of quantum theory is that the observer cannot be separated from the observed. Both are needed to "create" what we commonly call reality. If one is missing reality vanishes. In scientific jargon we would say that without an observer there is no reduction of the wave packet. Without an observer no reality can be created; matter remains just waves, waves of probabilities.

According to the physicist David Bohm, our perception of the world is the result of extensive conditioning of our brain through the ages.[1] This conditioning has created a separation—which he considers artificial—between humanity and nature and between human and human. In other words, for Bohm, our perception is responsible for the fragmentation of our universe. He believes that quantum theory implies that this conception is unsustainable, and that the world must be conceived as an undivided whole in which the observer and the observed are one. In this unity he includes not only matter, but also mind. For him, mind and matter are two aspects of the same entity. Without going too far, too fast, could we not imagine, for a start, that emotions are a bridge or an interface between our body and our mind?

After the radio, which translates electromagnetic waves into sound, and the fax, which translates electromagnetic waves into two-dimensional images, came the invention of the hologram, which translates electromagnetic waves into three-dimensional images. Now there are machines that are able to further translate electromagnetic waves into three-dimensional images that can even be "touched." Through an intimate interaction with a computer an imaginary environment can be created for an observer, who can then experience a virtual reality. In the most successful virtual environments, users feel that they are truly present in the simulated world. This simulated world "touches" them.

Could our brain, this very complex network of neurons, be such a machine too? A machine that creates, through its interaction with matter waves, a three-dimensional picture with shapes, textures, colors, sounds, smells, and tastes? A device through which wave packets collapse? A device through which one of the many possibilities contained in these waves packets becomes real, at least for us? Who chooses? Bohm said that we are all observers who have created the reality we live in. One of the questions we could ask ourselves is whether the world that we might be creating at every instant is real or virtual. Because most of us have the same brain, we all seem to create a similar reality; we could therefore say that our world is real, because it is the "same" for most of us. However, it is most probable that a human being with a damaged brain would create/experience a different reality. Is his or her reality less "real" than ours?

Another relevant question for us is: Does this world that our mind might create from matter waves also include our inside world—this world we cannot share but which is certainly not less real for us than the outside world? Did we create it? How real is it? How definite is it? Can we change it? Can our molecules affect it?

From the beginning of Christianity and during the following centuries, the Western concept was that our planet Earth was a static sphere around which inaccessible planets revolved in perfect circles, the whole thing covered with an immutable vault on which the stars, not less immutable, were hung like pictures on a wall. The human race was imagined to be a foreign and ephemeral phenomenon in a perfect and eternal world. Only in the sixteenth century did the observations made by Copernicus (1472–1543) and confirmed later by Galileo (1564–1642) reveal a different reality. The contemporaries of Copernicus and Galileo saw the earth move; the stellar ceiling that, until then, had looked rigid and impenetrable, collapsed.

Even though the earth had always been in motion and even though there never was a stellar ceiling, strangely enough this discovery was experienced as though it were a real cosmic event. As if their image of the sky could affect their whole being, people seemed, once the doors of the sky opened, to exit a prison where only their imagination had confined them, by general consensus. They felt free, reborn. Fruits of this new creativity could be seen in all fields of culture: religion, philosophy, art, literature, science, and technology. Modern science was born. This episode of human history could be a good example of the often underestimated power of imagination and its conditioning.

Quantum theory was born only a few decades ago. It has hardly traveled beyond the doors of research institutes and hardly started to infiltrate in the public mind. This theory could launch the next big scientific revolution, perhaps even more earth-shaking than the Copernican revolution. This time it is not the structure of the skies that collapses, but the very substance of the universe, and along with it that of our own flesh. After Copernicus and Galileo, we had to destroy, not without pain, the wall that consensus had placed between us and the sky. Could there also be a wall between matter and mind, placed there solely by consensus? At the borders of science, where theories stumble and speculations begin, can we find a breach?

EPILOGUE

In this immeasurable universe, of which we know so little, at the mercy of all possible and conceivable plagues since the day of our birth, our end is our only certainty.

Navigating the seas of reality

Thrown without our consent into an unlimited molecular network in constant mutation, miniscule ingredients drowned in an enormous soup heated by a mysterious fire and stirred by the laws of thermodynamics as a ladle, condemned sooner or later to dissolve, what power do we have? What is left for us to decide, other than to propel our boat toward the vortex that sooner or later will swallow us up, or in the opposite direction, where a wave could, for one more moment, bear us away toward life that breeds life?

NOTES

INTRODUCTION

1. Alexander I. Oparin, *The Origin of Life* (New York: Dover, 1953).

CHAPTER 7. THE SECOND LAW: CHAOS RULES

1. Peter W. Atkins, *The 2nd Law* (New York: Scientific American Books, 1984).

CHAPTER 11. FAR FROM THE CROWD

1. Ilya Prigogine and Isabelle Stengers, *Order out of Chaos: Man's New Dialogue with Nature* (Toronto: Bantam Books, 1984), 148.
2. Ibid., 171.

CHAPTER 12. THE QUANTUM REVOLUTION: MATTER'S DOUBLE NATURE

1. Fred A. Wolf, *Taking the Quantum Leap: The New Physics for Non Scientists* (San Francisco: Harper & Row, 1981).

CHAPTER 13. COMMUNICATION BETWEEN MOLECULES

1. Primo Levi, *The Periodic Table* (New York: Schocken Books, 1984), 232.
2. Erich Fromm, *The Art of Loving* (New York: Harper, 1956).

CHAPTER 14. THE HOLOGRAPHIC UNIVERSE

1. Olaf Nairz, Markus Arndt, and Anton Zeilinger, "Quantum Interference Experiments with Large Molecules," *American Journal of Physics* 71, no. 4 (2003): 319–25.
2. Martin Enserink, "French Nobelist Escapes 'Intellectual Terror' to Pursue Radical Ideas in China," *Science* 330 (Decembre 2010): 1732.
3. Sason Shaik, "Chemistry—A Central Pillar of Human Culture," *Angewandte Chemie International Edition* 42, no. 28 (2003): 3208–15.

CHAPTER 15. MOLECULAR MOODS

1. Roger Penrose, *The Large, the Small and the Human Mind* (Cambridge: Cambridge University Press, 1997).

CHAPTER 16. CHOICE: WHAT INSTRUMENT DO YOU PLAY?

1. Michael Bond, "Three Degrees of Contagion," *New Scientist* 2687 (2009): 24–27.

CHAPTER 17. MIND AND MATTER

1. David Bohm, *Wholeness and the Implicate Order* (London: Routledge and K. Paul, 1980).

INDEX

Page numbers in *italics* refer to illustrations.

BOOKS OF RELATED INTEREST

Morphic Resonance
The Nature of Formative Causation
by Rupert Sheldrake

Science and the Akashic Field
An Integral Theory of Everything
by Ervin Laszlo

The New Science and Spirituality Reader
Edited by Ervin Laszlo and Kingsley L. Dennis

The Biology of Transcendence
A Blueprint of the Human Spirit
by Joseph Chilton Pearce

Science and the Near-Death Experience
How Consciousness Survives Death
by Chris Carter

Darwin's Unfinished Business
The Self-Organizing Intelligence of Nature
by Simon G. Powell

Transcending the Speed of Light
Consciousness, Quantum Physics, and the Fifth Dimension
by Marc Seifer, Ph.D.

Where Does Mind End?
A Radical History of Consciousness and the Awakened Self
by Marc Seifer, Ph.D.